Otto Hahn

Springer Science+Business Media, LLC

Otto Hahn

Achievement and Responsibility

Klaus Hoffmann

Translated by J. Michael Cole

With 42 Figures

 Springer

Library of Congress Cataloging-in-Publication Data
Hoffmann, Klaus, Dr. rer. nat.
  [Otto Hahn. English]
  Otto Hahn : achievement and responsibility / Klaus Hoffmann.
    p. cm.
  Includes bibliographical references and index.
  ISBN 978-1-4612-6513-9   ISBN 978-1-4613-0101-1 (eBook)
  DOI 10.1007/978-1-4613-0101-1
  1. Hahn, Otto, 1879–1968. 2. Chemists—Germany—Biography. 3. Atomic
bomb—Germany—History. I. Title.
QD22.H2 H6413   2001
539.7′092—dc21   00-040045

Printed on acid-free paper.

Production managed by Allan Abrams; manufacturing supervised by Jacqui Ashri.
Typeset by TeXniques, Inc., Boston, MA.

9 8 7 6 5 4 3 2 1

SPIN 10770136

Translator's Preface

It has been a pleasure to translate Klaus Hoffmann's biography of Otto Hahn, finding many an unexpected turn of phrase and an elegant style missed in the verbiage of much contemporary writing in the general arena of science.

Otto Hahn is seen to be one who felt a piercing sense of guilt for his act of discovering nuclear fission when he first learned of the horrendous effect of the first and second atom bombs dropped on Japan. A period of mental conflict lasted for some significant time before being resolved by a gradual awareness that scientific discoveries are in some degree morally neutral whereas uses to which they are put may not be. Of course, he had the conflict of knowing that he was closely involved with colleagues, fellow countrymen, who were working on the possibility of creating an atom bomb during the course of the Second World War, even though it became apparent that the project would not come near succeeding.

The translator has found a number of differences between the author's account of various matters and official records of recognised conditions. At the time when the original text was written the transcripts of the secretly made recordings of the German scientists when detained in England had not been released; they were not released until early 1992 at the Public Record Office, Kew, and thus were unavailable to the author. In 1993 the complete transcripts of those conversations were published in *Operation Epsilon: The Farm Hall Transcripts* by Institute of Physics Publishing, together with an Introduction and Archival Note by Sir Charles Frank, one of the few visitors to the group during their detention.

Notes on these differences and one or two other points are set at the end of this Translator's Preface.

The translator is of the view that there are occasions in which the essence of a text is destroyed if it is translated in too broad a manner, just to satisfy the mere dictate of the fashion that asks for undemanding contemporary colloquial usage. Otto Hahn was of the generation brought up during the last two decades of the nineteenth century, and he learned his English in the Edwardian era at the hand of others of an even earlier generation. The natural flow of the author's text translates well into such a usage of English, and that flavour has been deliberately kept. Doing so has the advantage that the author's excursions into amusing, mocking, ironic, sardonic, or even laconic turns of phrase stand out even better.

I must extend my thanks to the author and to Thomas von Foerster of Springer–Verlag for their help in elucidating a number of points in the translation. But despite all efforts to remove errors in any translation, one becomes aware that some slip through all the processes of checking. For such instances the translator begs indulgence. He will be satisfied if the reader finds the book as fascinating and compelling as he found the original.

Leyburn, England, UK   J. MICHAEL COLE
December 1999

Contents

1
Atom Bomb and Nobel Prize

1.1 Hidden Microphones

In a lonely area some twenty five miles west of the old university town of Cambridge, there lies on the edge of the small town of Godmanchester the country seat of Farm Hall. One Judge Clark had had it built for him in 1728. In the dawn of antiquity the region was part of the Province of Britannia.

Judge Clark, a passionate archaeologist, knew that and began, foot by foot, to scour his estate for antiquarian objects. To protect himself from the gaze of inquisitive passers by he had a high wall built around his property. But his hope of making Farm Hall famous in his lifetime through its antiquarian finds was never to be fulfilled. It was not until at least two hundred years later that the world was to learn of Farm Hall—as a country seat turned into an exclusive prison.

In July 1945 a group of internees was brought there under armed escort by the British military, "10 persons and 4 ordinary prisoners of war" as the register's entry has it. Of the red brick, three storey house, with its many windows, a few of which were bricked up[1], the prisoners could only snatch

---

[1] *Translator's Note:* In 1695 a 'window tax' was introduced, levied on all windows above six in number; it resulted in many buildings having some windows bricked up in order to reduce liability to the tax. The law was eventually repealed in 1851.

a fleeting impression as the armed soldiers hurried them into the house. Here a sergeant divided them into groups of two men each.

When the first two men had shut the door behind them they carefully inspected the room, the windows of which—to their surprise—were not criss-crossed with barbed wire. They glanced back into the garden which gave a friendly impression, even if one of being overgrown. A wall about as high as a man, here and there luxuriantly overgrown with ivy, drew itself around the property. Soldiers were on guard patrolling around the house, and these two remembered again their desperate situation. One began to tap the walls suspiciously.

"I wonder if there are any microphones built into these walls?"

"Microphones in the walls?" The one spoken to laughed. "Oh no, they are not so cunning. I don't believe that they know of such Gestapo methods ... ". The pair belonged to a group of Germans who for some time had been in the safekeeping of the British Secret Service. Who were these people who were of such special interest to the Secret Intelligence Service?

For the high ranking forces' officers there could be no question about it. Their behaviour had absolutely nothing military about it. In general their detention gave little occasion for concern and made the security precautions appear superfluous. The watches of the guard wondered at the way in which the prisoners almost every day passed the time in extended discussions and in scribbling incomprehensible formulae and strings of symbols. A few soldiers understood some German, but of the shreds of conversation that now and then they could pick out they could make no sense.

Amongst the internees there were ten scientists—physicists and chemists —some academics of international distinction: Walther Gerlach, Otto Hahn, Werner Heisenberg, Max von Laue, and Carl Friedrich von Weizsäcker. In addition there were Erich Bagge, Kurt Diebner, Paul Harteck, Horst Korsching, and Karl Wirtz. They were specialists in the area of atomic research and played a leading role in the secret uranium programme of the German armed forces. In that research project the Germans had pursued the goal of making the principle of nuclear fission, discovered at the end of 1938 by Otto Hahn and Fritz Strassmann, into a useable technology for warfare.

Gerlach had ended up as the Director of the Nuclear Physics Working Party in the Reich's Research Council of the Chief of the German Atomic Project. Heisenberg and von Weizsäcker were there as leading theoreticians. The group of Diebner and Bagge belonged to a competing undertaking.

The way in which and the means by which these scientists were seized was up to then without parallel in history. An American commando unit had tracked down the atomic researchers in southern Germany shortly before the end of the Second World War and had taken them prisoner. Important evidences of their work were seized, and their experimental plants dismantled or destroyed.

In strictest secrecy and under the strongest possible escort the ten scientists were brought first to France and Belgium, and finally to Britain, to the isolated country seat of Farm Hall. The operation was carried out under the codename *Epsilon*. It was only later that they were to learn that their present home served as a training centre of the British Secret Service for special agents. At first the scientists had no presentiments of the mysteries that Farm Hall's walls hid, nor that they would thoroughly inspect the interior of that selfsame house. Bagge noted in his diary, "It is remarkable how many, many paintings there are on the walls inside".

The worry about their own fate and the likewise unknown fate of their families weighed heavily upon the internees. The days passed frighteningly slowly. They occupied themselves with exchanges of scientific opinion, working in the garden, and a little sport. Otto Hahn was proud that he, at sixty six, set a 'house record' for the ten kilometre race. This peaceable life could not, however, escape the tensions within it that sprang from their different political persuasions, nor from a certain scientific rivalry. Of Otto Hahn and Max von Laue it was known that they had never sympathised with National Socialism, nor yet had been responsible for Hitler's collapse.

6 August 1945 seemed to be like any other day. Nothing pointed to any change in the monotonous run of daily life. This abruptly altered when a courier from London unexpectedly arrived and handed an urgent order to Major Terence Rittner, who was responsible for guarding the Germans.

Soon brisk activity prevailed within the house. The Germans had to leave their rooms. Specialists in the Secret Service set to work on thoroughly overhauling the listening system. Their special interest lay in the tiny microphones hidden behind the numerous paintings. They were very relieved that the Germans had evidently not discovered the microphones.

The order which London Centre of the Secret Service had conveyed was mysterious even to Major Rittner. On that day, 6 August, the prisoners were to listen to the BBC News. Everything that they were to say and discuss about it had to be listened to over the microphones, noted down in its entirety, and be immediately sent back to London. That their conversations had to be harvested was nothing new. But to Rittner the order about the BBC broadcast seemed beyond explanation.

The BBC's Six O'Clock News consisted of only a very few sentences, "... President Truman has made known a great achievement of the Allies' scientists: they have created the Atom Bomb. The first has just been dropped on a Japanese army base and had as much explosive power as two thousand of our ten ton bombs". This announcement brought loud cheers from the British officers and soldiers. Now it was completely clear why 'the Germans' had had to hear the Six O'Clock News. Major Rittner's entry in the register[2] for 6 August 1945, states, "Shortly before dinner I informed Professor Hahn of the BBC's announcement about the dropping of the

---

[2] *Translator's Note:* See, 'Item 1. II. 6 August 1945. Farm Hall Report No. 4'. (1993) *Operation Epsilon* (Institute of Physics Publishing: Bristol, UK, and Philadelphia, PA), p. 70.

atom bomb. From that moment Hahn was very distracted and declared that he felt himself personally responsible for the deaths of the hundreds of thousands of people, because his discovery had made the building of the bomb possible. He told me that his first thought had been to contemplate suicide when the frightful potential of his discovery had been demonstrated. After a considerable quantity of alcohol Hahn regained his composure and we went down to dinner, where he delivered the news to the other guests".

The notes in Otto Hahn's own diary reveal not only his perplexity, but also his quiet hope that the announcement about the atom bomb might not be true[3]: ... *I want it not to be true, but the Major assured me that it was no reporter's fanciful tale but an official announcement of the President of the United States. I almost fell to pieces at the thought of this new, great misery ... .*

As the tape recordings reveal, they discussed the news agitatedly and with violent emotion. Gradually, however, doubt began to settle upon them.

Heisenberg turned to Otto Hahn, "Was the word 'uranium' used in the context of this bomb?"

"No".

"Then it has nothing to do with atoms ... I rather believe that it is a high pressure bomb, and that it has nothing at all to do with uranium. It will be a chemical device in which they have enormously increased the explosive power ... ".

Heisenberg attempted to leave his thoughts at that point at which it was evident that a new explosive might be manufactured from atomic hydrogen and oxygen. But Harteck, who was following these explanations shaking his head, would have none of it. On its own, under the most favourable circumstances, so he estimated, such a chemical explosive material could have only ten times the effect of the most powerful bomb known. The atom bomb, however, should exhibit an explosive yield of 20,000 tons of TNT. Therefore only uranium could be in question as the explosive material of the bomb. When the cue 'uranium bomb' popped out Otto Hahn could no longer hide his inner agitation.

Gradually the discussion dried up. They would wait for further news. Bagge took up his diary and wrote, "Herr Hahn very agitated, so hopes Heisenberg right, because he is frightened by the thought that his own discovery could have significant consequences for war".

1.2   Death Clouds over Hiroshima

In the early morning of 6 August 1945, a solitary aeroplane flew over the Japanese city of Hiroshima at a very high altitude. The morning sun flooded

---

[3] Otto Hahn's authentic statements are printed in italics.

down upon the town of more than 300,000 inhabitants that lies gracefully in the plain of the Ota delta surrounded by ranges of hills, a few of which stretched into the town.

Up to then Hiroshima had been spared from American bombing raids. The single American B29 bomber which droned along at about 33,000 feet this morning frightened nobody. No citizen thought that this B29 would rain death and destruction over the entire city.

A few minutes after 8 o'clock, at the command of the captain, Colonel Paul Tibbets, Major Thomas Ferebee, the bomb aimer of the B29, released their burden of death. The bomb on its parachute slowly approached the city centre . . . .

8.15 a.m., local time. The townsfolk went about their daily work. Many people poured into the city every day from the surrounding villages to earn their daily bread. Business had already opened. The children gathered in the school yards in order to go in to their classrooms together. 8.15 a.m. on a sunny summer's morning.

At the planned height of between 1,600 and 2,000 feet the bomb went off. An inferno broke loose. The explosion surpassed anything there had ever been. The flash, which bathed the mile wide area in a scorching light, was followed by a fireball of gigantic size that spread outwards at break-neck speed. A huge cloud boiled and seethed upward to over 50,000 feet, which finally turned into a giant mushroom, the foot of which must have been three miles in diameter. A long, continuous, dreadful thunder, which nobody had ever heard before, accompanied the awful spectacle.

After dropping its cargo of death the B29, with engines roaring, turned off in a nose dive and had already gotten itself some ten miles away when the bomb detonated. In spite of all the precautionary measures, the crew were taken by surprise by the power of the explosion. Just as under heavy anti-aircraft fire the B29 rolled from side to side. Even inside the hermetically sealed cabin they felt the blasts of heat and pressure.

"I've never seen anything like it—not ever", said the co-pilot, Captain Robert A. Lewis, at the debriefing. "The city looked as if it had been torn in pieces". And the skipper, Captain William S. Parsons, who had become pointed during the flight, commented, "The Japanese will think that a meteor has hit their city".

The temperature of the bomb's explosion was estimated at ten million degrees, a magnitude which occurs only inside the Sun. In the immediate vicinity of the centre of the explosion, the so called 'ground zero', within a circle of about a quarter of a mile in diameter, the molten stone and brick of houses had been reduced to a glassy mixture by the searing heat. Of the people who had been there no trace was to be found. Only on the steps of a building were their footsteps found or their shadows burnt into the remains of walls. The destruction was unimaginable. Huge buildings had collapsed like houses of cards. Pieces of walls, like projectiles, flew through the streets. A fire storm followed the blast. The city burnt all day. The embers glowed for a week. Indescribable scenes were played out amongst the population.

Whoever escaped the heat and blast waves became a sacrifice to the radioactive rays, which became a new category of cause of death. Many who had survived the first inferno suffered the insidious radiation sickness, often after a long, agonisingly painful fight against death. In 1945 the people of Hiroshima mourned 141,000 sacrificed, and in 1946 there were another 10,000. Today the grim reaper scythes his yearly harvest amidst the Japanese, who on 6 August 1945 had stopped in Hiroshima and had been exposed to the death-bringing radiation. But that did not draw a line under the balance sheet of horrors. The descendants, also, of each unlucky person suffered and suffer from deformities as a consequence of radiation damage.

The dropping of the atom bomb was a crime. American politicians and military had had no conscience about the death of the civilian population of a great city in their calculations of their power politics. For the deployment of the bomb there existed, looking at it from the military point of view, no necessity whatsoever.

Germany, for which the bomb was originally intended, had already capitulated on the 8 May 1945. Japan was at the end of its military strength, and the entry of the Soviet Union into the war would have had to have led to the speedy capitulation of Japan. An atom bomb had not been needed to bring it about.

1.3   Hot Tempered Discussions

In Farm Hall all ten scientists crowded round the radio in order to listen to the Nine O'Clock News. They could hardly contain their excitement and seemed to be entranced by the set. Otto Hahn sat away from them, all quiet, his head bowed.

"Here is the News. It is dominated by a tremendous achievement of Allied scientists—the production of the atomic bomb. One has already been dropped on a Japanese army base. ...reconnaissance aircraft could not see anything hours later because of the tremendous pall of smoke and dust that was still obscuring the city of once over three hundred thousand inhabitants... ." The announcer named read out details. The atomic bomb project had cost over 500 million pounds sterling (2 billion dollars). At times up to 125,000 people had worked on building the bomb.

But the German atomic scientists who sat imprisoned in Farm Hall and listened to the news waited for quite different information. Then, with conflicting feelings, the expected keyword came; uranium had been used for making the atomic bomb. A uranium bomb had destroyed Hiroshima.

The emotions of the scientists after this feared news can be written about only with difficulty: paralysing horror on the one hand and anger and annoyance on the other. Then by and by a sensible discussion came into being, which took on a passionate form. With the help of the concealed microphones these conversations were recorded.

"I think it is dreadful of the Americans to have done it", said von Weizsäcker with indignation. "I think it is madness on their part".

"One could equally well say that is the quickest way to end the war", added Heisenberg. He lamented that the German atomic research had not been pushed ahead to the same degree as the V1 and V2 weapons[4]. But then Heisenberg had to admit, "We did not have the moral courage to recommend the 1942 government to put a hundred and twenty thousand people on the job".

An interjection of von Weizsäcker was to give the conversation a new turn: "I think that the reason we did not do it was that, fundamentally, not everybody wanted to do it. If everybody had wanted Germany to win the war, then we would have been able to build it".

At that, Hahn said, *I am glad that we did not build it.*

Later, when the first shock had soaked into quiet reflection Hahn entrusted his thoughts to his diary. At the end of his entries for "The Day of the Uranium Bomb", it says, *I am now glad that we had no way or means of developing a bomb for if it had been possible to build it in Germany during the war one would have been compelled to use it against England. To me that is unthinkable. I do not begrudge the Americans the fame.*

Not everyone thought so. A few of the interned atomic researchers in Farm Hall had, at the time, driven by ambition, given themselves to the German atomic project, and now could not conceal their disappointment. Bagge made himself their representative when he irritably said in reply, "I find it absurd of von Weizsäcker to say that he had wished the enterprise no success. That might have been true of him, but not for all of us". Diebner livelily agreed with him. Suddenly the earlier rivalry was awakened again, the old mistrusts glinted with life. In front of all the younger scientists like Bagge, Diebner, and Harteck, Wirtz put forward the unrealistic opinion that with a German atomic bomb as a bargaining tool it would have been possible at the end to have negotiated more favourable conditions for surrender[5].

Gerlach thought the same. In general he had a very depressing effect, rather like a defeated general. Hahn sensed it and took his friend to his side to soothe him[6]. But he could only do a little. Gerlach was afraid that they would not be safe to return home for the rest of their lives. According to his conviction, in every case they were 'saboteurs' or 'traitors', either because they had not pushed ahead with the production of their own atomic bomb

---

[4]See Note 1 in Translator's Notes on the Text.

[5]See Note 2 in Translator's Notes on the Text.

[6]See Note 3 in Translator's Notes on the Text.

with sufficient effort or because they had finally 'worked together with the Allies'. In a foreign country they were now seen as 'war criminals'[7]. How was one to break out of this vicious circle of accusation?

Disconnected shorthand notes in Gerlach's hand give evidence of his perplexity and confusion, "The whole work during the war was for nothing ... Must each task that brings men help bring also simultaneously their destruction ... ? The situation in our circle is ever more difficult and tense. One comes across peculiar opinions ... Very great disappointment about the attitude of a few gentlemen ... "

Hahn took no further part in the later excited discussions. He obviously suffered a great physical shock. The others noticed what was wrong with their senior. "Keep your eye on Hahn" they whispered to each other, for they feared that he might take his life[8].

The pairs of scientists sat up into the early hours having discussions. Again and again each and every one of them talked about the same questions.

"How did they make it?", Heisenberg repeated time and time again. "It was a disgrace that we who worked on this project matter could not at least find something, as it has actually been made ... ", Gerlach shook his head again and again. He could not sort things out in his mind about the American bomb; not on this day he couldn't.

Max von Laue philosophised out loud. "When I was young I wanted to pursue physics and experience world history. Physics I have pursued, and I have witnessed world history—in all truth, I can now say in my old age that I have done both".

At three in the morning when the others had finally gone to bed, von Laue knocked on Bagge's door[9]. "We must do something. I am very worried about Hahn", he whispered. "The news this evening shook him. I fear the worst".

Both stole to his room and cautiously opened the door. Otto Hahn twisted and turned on his bed sleeplessly. For a little time both kept watch from the door. Once they saw that he was quiet and had fallen into a deep sleep they went back to their room. "The pitiful Professor Hahn", Bagge wrote in his diary before he laid himself down to rest... .

So much for the occurrences in Farm Hall on 6 August 1945. With the help of diary notes and the now well known tape recordings obtained they can be faithfully reported. The transcripts of these recordings were only

---

[7]See Note 4 in Translator's Notes on the Text.
[8]See Note 5 in Translator's Notes on the Text.
[9]See Note 6 in Translator's Notes on the Text.

made available by the British Secret Service[10] in 1992 after protests[11] from scientists that these documents should be reclaimed for historical research were eventually successful. Fragmentary extracts of these bugging minutes had already been successfully made available earlier in various ways.

The next few days are completely recorded in further reports in the press and radio reports about the American atomic bomb. On 9 August, two days after his return from Potsdam, President Truman gave his press conference. The representatives of numerous press agencies and newspapers, and radio reporters gathered in the White House. Everyone hoped to hear at first hand of the results of the Potsdam Conference. They also wanted to learn about Truman's ideas about the future collaboration of the Allies, especially with the USSR.

It was with astonishment, however, that the journalists noted that Truman's self-satisfied speech was weighed down with threatening undertones. The US President made play of the new factor in the power of the United States, the atom bomb. Its creation had cost the country over two billion dollars and must be recorded as the biggest scientific gamble in history. The USA had won this game brilliantly and through doing so had become the most powerful country in the world.

The prisoners in Farm Hall also heard Truman's speech. Otto Hahn and a few others became conscious from this speech of the danger that could arise from the atom bomb being in the hands of politicians, fears which not all in the circle of scientists shared. A discussion about it did not get off the ground because Truman's further revelations about the history of the uranium bomb claimed their undivided attention.

"Already before 1939 scientists were convinced that it was theoretically possible to release the energy of the atom. But nobody knew that could be managed to be done practically. In 1942 we discovered that the Germans were working feverishly to make atomic energy available for their war machine, by which they wished to enslave the world. British and American scientists have in a joint endeavour undertaken a race of discovery with the Germans. We have won the battle in the laboratories just as we have won the battle in the field".

Germany had surrendered. Japan, however, turned down the ultimatum and took the battle further. Truman also spoke about that in his speech, "If the Japanese emperor does not now accept our conditions then they

---

[10] *Translator's Note:* The transcripts were actually made available on 14 February 1992, for public inspection by the Public Record Office at Kew, Class WP 208, Piece No. 5019; *see* the reply dated 13 January 1992, to the President of The Royal Society from the Lord Chancellor, *Operation Epsilon*, p. 16.

[11] There was, in fact, a formal request from the Presidents of The Royal Society and The British Academy, which was also signed by thirteen other leading academic scientists and historians, *see Operation Epsilon*, pp. 14–15.

must expect a storm of destruction such as has never been known on this Earth up to now".

On the same day the second American atomic bomb was detonated over a Japanese city—over Nagasaki.

1.4   Rumours about Professor Hahn

The world was still reeling from the shock of the devastating effects of the atomic weapon when the first press reports were beginning to introduce the 'inventor' of the super-bomb. A number of important scientists, so it was said at first, had been working together on the building of the bomb. Then the first names were named.

The Number 1 of the *Allgemeine Zeitung* of 8 August 1945, published in Münich by the US Army, was devoted to the dropping of the atomic bomb, as were its headlines. Under another headline they published a statement of the physicist Wilhelm Westphal, "Although more detailed news is still lacking, there can never be any doubt that the atomic bomb depends upon the fission of uranium atoms when bombarded with neutrons, discovered by Hahn and Strassmann in 1938".

The name of another worker at Otto Hahn's institute was also named in connection with the development of the atomic bomb, that of Lise Meitner. She was even spoken of as having played a crucial part. The sensational press intimated that Lise Meitner, the 'Jewish assistant of Professor Hahn', had fled abroad with the secret of the production of the bomb. All documents about the 'Hahn links' were deposited by her, so it was said, in a steel box of a bank in Vienna during her escape and from there were handed over to the Allies.

According to other information the Hiroshima atom bomb had been nothing other than an American built one of Hitler's V3 weapons, and thus a German invention. Many another fable was heard in later years. Twenty years after these events it was reported in the Madrid newspaper *Pueblo* on 6 August 1965, in large headlines, "La Bomba Atomica era alemana" (the atomic bomb was German).

The name of the German atomic scientist Otto Hahn appeared more and more frequently in the newspapers of 1945. The reports came thick and fast that Hahn was in the USA. He had then been seen in the notorious atomic cities. None other than he had given away the secret of the atomic bomb to the Allies.

There were also rumours in circulation about the hazardous route by which Otto Hahn had at that time come to the USA. A thirty strong special commando unit was parachuted behind the enemy lines in the middle of the war and had abducted the academics. That, at any rate, was reported in all seriousness years later by the Honourable Member Stringfellow of the United States House of Representatives about these 'heroic' undertakings he would have wished personally to have led. According to another account Otto Hahn, in concert with specialist colleagues, was supposed to have consciously hindered the work on the German atomic bomb. In *New Statesman and Nation* it was said that it was hoped that Otto Hahn would be awarded the Nobel Prize for Chemistry, and the Nobel Peace Prize as well, for his contributions, because he had known the secret of the making of the atomic bomb but had not betrayed it to Hitler.

Certain circles, who after the fall of the 'Third Reich' laid to heart the revival once again of the old myth of the stab in the back, later repeated these and similar accusations. In 1955 one of the German-American newspapers published in the USA described the German atomic scientists as miserable traitors, with Otto Hahn at the head, because they had withheld from their country the atomic weapon that would have brought victory.

Was it true or was it just a story?

Documents confiscated by the western Allies, in which Hahn's name was mentioned repeatedly in connection with the German atomic project, appeared to confirm that the academic had collaborated on this particular project. Was Hahn therefore, nevertheless, supposed to have known the 'secret' of the atomic bomb?

1.5 Criminal Investigation for a Nobel Prize Winner

On Friday, 16 November 1945, they were sitting together, early, as usual, to study the latest press reports. Professor Hahn had just gotten himself comfortable in an armchair and like the others was leafing through the just delivered English daily papers when he was disturbed. Heisenberg referred to a short article in *The Daily Telegraph* which said that Otto Hahn was to receive the 1944 Nobel Prize for Chemistry. There was understandable excitement. So the Nobel Prize in the natural sciences was awarded only for outstanding scientific achievements. Of the circle of scientists in Farm Hall the physicists Max von Laue and Werner Heisenberg had received this much sought after honour. During the Nazi era Otto Hahn had been talked of simply as a 'clandestine Nobel Prize winner'.

Modestly Hahn, who was heartily congratulated by everyone, doubted that the announcement was authentic. But in the next few days the Swedish news agency confirmed the announcement. Hahn would receive the Nobel Prize for his scientific work on the discovery of the fission of the uranium nucleus. Unfortunately it was not known at that present time where the distinguished academic was. For a long time he had been missing.

With considerable delay the first post from Sweden reached the candidate. The first thing in the letter was that Hahn might kindly request to not be invited to Stockholm 'because of the disquiet about the atomic bomb'.

However, there then followed the official invitation to the celebration taking place from 10–12 December 1945, on the occasion of the bestowal of the Nobel Prize. Professor Hahn was thrust once again into the centre of 'Allied Interests'. American and British officers endeavoured to convince him that under the present circumstances on no account could he receive the Nobel Prize in Stockholm in person.

Hahn was of another opinion and demanded that it would at least have to be announced why he was not allowed to travel to Sweden. Then his written reply would not be conveyed, said the military in reply. Further objections were rejected with the comment "You are German, you have lost the war".

In his will Alfred Nobel had stated that every year those scientists and personalities "who in the previous year have brought to mankind the greatest benefit" should be awarded a prize. The atomic bomb, that frightful spectre over mankind, was undeniably a consequence of the discovery of Otto Hahn. A Nobel Prize for the atomic bomb as well? Much to his regret, all his life Otto Hahn had been confronted with such contorted notions. He was hit particularly hard by the publicly stated reproach that the Nobel Prize was 'thirty pieces of silver' for the bomb that he brought the Americans and had devastated Hiroshima and Nagasaki. Until then there had been no comparable situation in which a scientist had come to sense the tragic consequences of his discovery so mercilessly. The Nobel Prize which Hahn was to receive in recognition of his scientific work on the discovery of nuclear fission must have made this seem an irony in view of the suffering of Hiroshima and Nagasaki. These smouldering conflicts of conscience, however, also helped Otto Hahn to escape to a new view and insight. At this time there matured in him the decision to concentrate in future on his strengths, to exclude the misuse of scientific investigations, and to appeal to the responsibility of scientists. All his life Otto Hahn had, admittedly, not been able to escape reproach and self-accusation; he had, as the discoverer of the fission of uranium, also to personally bear part of the responsibility for the atomic bomb. From that position he was also not freed from the gradually evolving view that his discovery had indeed been the means of making this means of mass destruction possible; but at the same time also the realisation was growing that it could be developed into a new source of energy for the benefit of mankind.

A decade after Hiroshima, in 1955, Hahn lamented amidst his circle of friends, *People say that I am guilty of there being the atom bomb and of Japan having had to suffer so terribly, and that our future has been endangered. But I am the only one who has done my scientific duty.*

If one wants to weigh rightly the tragedy that lies buried in these words, one has to get a grasp of the background and context of the role they played in the life of this scientist. Was it inevitable that Hahn's discovery had to lead to the atomic bomb? Should he really have withheld the secret of the bomb from the Nazis in order to give it away to the Allies?

Otto Hahn had always very definitely rejected the assertion that his life had been unusual and full of dramatic peaks. Himself his greatest work he assessed only to be *good scientific work*, and no more. In these words Hahn's personal modesty was brought out. Nevertheless, there existed no doubt that the life of this important scholar unwound as a procession of quite out of the ordinary events in dramatic times.

2

Boyhood, Studies, and the First Probationary Years

2.1   The 'Good Year' 1879

"The year eighteen hundred and seventy nine is a good year", declared Carl Ramsauer with glass in his raised hand during a birthday party. Anyone who knew him better immediately knew that Ramsauer was not referring to the wine alone. He was well known to his physicist guests for his jocularity. Not only did he belong to this year, but also Otto Hahn, Albert Einstein, and Max von Laue.

These remarks Otto Hahn had liked to use and—following a suggestion of Max Planck—to include his colleague Lise Meitner, who had been born just beforehand in November 1878. "That is of no consequence", averred Hahn, "Girls have always been a bit forward".

Otto Hahn came from Frankfurt am Main. His parents, however, were not established citizens of Frankfurt. His father, Heinrich Hahn, who had learnt the glazier's craft, had earlier first settled in the free city. There, in 1875, he married a young widow, Charlotte Stutzmann, née Giese, who had already taken on the care of a son, Karl, born in 1870, of her first husband. The married couple Hahn were given three sons, one after the other, Heiner in 1876, Julius in 1877, and on 8 March 1879, the youngest, Otto.

Helped by the industrial upswing which set in after winning the war of 1870–71 and the payment of reparations by France to Germany, the hard working and prudent father Hahn succeeded in building a secure existence. He expanded his modest handicraft business to a larger undertaking which was founded in 1836 as "Glasbau Hahn" and today is known world wide.

Father Hahn also engaged in local politics. As the representative of the upper artisan class he worked for a few years as a town councillor for the Democratic Party.

From the spring of 1885 Otto Hahn attended a secondary school in his home town. His brother Karl was a pupil at the Goethe Gymnasium, an apparently humanistic educational establishment. Otto Hahn was never entirely able to free himself from the feeling of having no real sense of supporting a humanistic outlook. *I am only a secondary school pupil* he used to say, even in his old age.

Notwithstanding the prosperity of their parents the brothers were brought up to be thrifty and modest, and happy, but otherwise with all the preferences of a carefree youth.

Otto Hahn remembered particularly clearly his childhood reading, in which he confessed to having completely devoured the fantasy tales of Cooper, Wörishoffer, Niemann, and Jules Verne. *We did not yet know Karl May*, he added a little regretfully. Later he counted popular technical-scientific books amongst his preferred reading.

I was a quite good pupil, but never an excellent one, I was never first, commented Otto Hahn about his school performance in a review in his old age. It is interesting how he assessed the teaching in the fields of natural science: *In spite of all his efforts the physics teacher did not succeed in making physics interesting for us. The chemistry lessons bored us to sleep, and yet I was increasingly interested in the subject.* The consequence—the school boy Hahn carried out his first experiments in the laundry room at home, his *chemistry games*, as he called them. In this way he learnt to produce hydrogen, to burn coal with oxygen, and took delight—definitely not harmlessly—in explosive reactions with sodium, phosphorus, and potassium chlorate.

At the wish of his father, who was successful in building and purchasing houses, Otto Hahn was to be an architect. But the son found that he had *no talent in drawing, no artistic gift*, therefore he was *unsuitable for becoming an architect*.

At Easter of 1897 Otto Hahn passed his matriculation. In the hearing of the professor of chemistry he later mockingly described his performance as *my matriculation certificate had three straight A-s, however, not in chemistry, mathematics, and physics, but in—gymnastics, singing, and religion.*

Otto Hahn's brothers also concluded their education successfully. Karl set out on a pedagogical career, and later spent a year as a graduate secondary school teacher in classics at the Goethe Gymnasium in Frankfurt am Main. Heiner took over his father's business, and Julius, who was allowed to follow his inclinations, became an art dealer.

Otto Hahn, who had latterly attended evening classes on organic solid fuels and had been able to convince his father that the natural sciences were his real calling, made up his mind to study chemistry.

2.2 University Studies in Marburg and München

When he was eighteen years old Otto Hahn left for the University of Marburg to register there. *Because it was then said that Marburg had no university, but was a university, I decided to go there*, was the reason he gave for his decision.

Hahn's student days were entirely carefree. Next to organising his studies and practical work, they were filled in good part with jokes and pranks. As a member of one of the then countless student societies he fought like the others and *quickly learnt the decent drinks*. Thus Hahn quite uninhibitedly described in his *Personal Confessions* how on a Sunday morning he would have to be gotten out from under a table with a broom. His parents not infrequently received postcards saying *I send you warm greetings from an enjoyable evening's drinking*. No wonder that to occasional questions about the well being of his youngest Father Hahn gave the general answer, "My son is at Marburg and drinks beer".

At the university the student Hahn enrolled in chemistry as his main subject with Professor Theodore Zincke. *The main lecture by Zincke was instructive, the presentation factual*, opined Hahn later. As his second subject he chose mineralogy, and crystallography, physics, and mathematics as his subsidiary subjects. But he was not best in physics. There were good reasons for that. The physics professor, Franz Melde, was an old man whom the students did not take seriously. He held his lectures early in the morning from eight to nine o'clock. That Melde started too early was soon the unanimous opinion of the young students in Marburg. The student Otto Hahn was for that reason seldom to be found in the physics lectures. As he freely confessed, to his regret he was never again able to make up for this omission.

The following episode has been handed down about this lack. Years later a scientific dispute between Professor Hahn and the physicist Lise Meitner was overheard by eavesdropping. The conversation took place in the stair well of his institute, and after a few technicalities on this and that there came the crushing words, "Hähnchen, go upstairs and do some chemistry, you understand nothing of physics".

Things seem to have been no better organised with mathematics at Marburg University. *Greatly missed was a short introductory course. In the first hour of the introduction to higher mathematics we young chemists understood not a word. We didn't go back.* Marburg at the time was an established German university which was very proud to count itself as the best in the state. It must have been thought that the cultivation of science in this seat of learning was in full flower.

Otto Hahn: *The end of the closing century was spent in the usual training course of young chemistry students who had no further ambition, and so made life as comfortable as possible for themselves. Not much notice was taken by us about the cultivation of science. Thus my time as a stu-*

dent passed carefree and with many cheerful hours and happy experiences;
carefree because I had never before become a scientist, and believed that for
a position in industry it was not necessary to be cultivated in more than
the principal subject, namely chemistry. Had I had any presentiment of my
later research then I would have given more time to my subsidiary subject
of physics, and to mathematics also.

After the end of the second year, Otto Hahn changed universities, as was
the custom of the time. In München he heard for a year the lectures of the well
known authorities in chemistry Adolf von Baeyer, to whom the synthesis of
the much sought after indigo belonged, and Karl A. Hofmann was among
his lecturers. *But my attendance at lectures in the fields of chemistry was*
not very regular.

In Marburg again, Chemistry Candidate Otto Hahn began his doctoral
work. Professor Zincke, amongst whose teachers the famous chemists Friedrich Wöhler and August von Kekulé could be numbered, gave him an exercise in preparational organic chemistry. *I bought myself a litre of isoeugenol*
which smelt much like oil of cloves, and made bromides, reported Otto
Hahn of his work. *A very beautiful crystalline derivative was obtained. It*
could be tested with the simplest of means. From this one learnt to record
well and to work conscientiously. I was very hard working, and worked in
the evenings at home as well.

By 8 July 1901, Otto Hahn had submitted an application for registration
for the oral part of the doctoral examination in the Faculty of Philosophy
of the university. Such a method of proceeding would take us by surprise
today for at this point the student Hahn had far from ended his doctoral
work. Professor Zincke endorsed in writing that the work would be ready in
the following year. "On bromine derivatives of isoeugenol", his dissertation
so named was handed in on 5 October 1901. Together with his student
friend Dahmer, he went into the examination on the 24 July. Today we are
rightly curious which facts were enquired of the examinees. In inorganic
chemistry Zincke asked both candidates about properties of neighbouring
groups of elements, lingering on bismuth in order soon to roam hither
and thither all through the then opening territory of organic chemistry:
diazo compounds—interesting for dyes, heterocyclic compounds—from the
pyroles to the tetrazines, indole and derivatives, and lastly synthetic indigo
and the medication antipyrine (phenazone).

In those days one did not need to know anything about radioactivity and
atomic theory in physics. Today it belongs to the basic knowledge of every
student of the natural sciences. Hahn and Dahmer were asked much more
about classical mechanics and thermodynamics: inertia, the motion of a
pendulum, the flow of heat in solid, liquid, and gaseous bodies, planetary
motion, gravitation, *etc.*.

"The answers of both candidates were, without exception, prompt and
correct, so that that there is nothing against them that should deny them
a rating of commendable", thought the mineralogy lecturer. "Both candid-

ates were so splendidly well versed that I am in agreement with the granting of good ratings to each", Zincke agreed with the report of his colleagues. With the appellation 'magna cum laude', Otto Hahn survived his doctoral *viva voce.*

On the occasion of the sixtieth anniversary of his doctorate in 1961, Otto Hahn received a congratulatory letter from the Heidelberg professor of chemistry Karl Freudenberg, who informed him that on that very day he was engaged in studying Hahn's doctoral work of 1901. The good crystalline bondings that Hahn had then found served Freudenberg as demonstration experiments. "You see that you have also been useful as an organic chemist", he wrote to Otto Hahn. The latter quick wittedly commented, *So perhaps I might have made something of myself if I had stuck to organic chemistry... .*

2.3   Soap Bubbles and Exploding Chlorine Gas: Dr. Hahn as an Assistant Lecturer

Before Dr. Phil. Otto Hahn looked for employment in industry he did his year's military service in the 81st Infantry Regiment in Frankfurt am Main. One year later a freshly promoted vice sergeant major left the regiment. Hahn did not seek to be a reserve officer, a rank much sought after out of snobbery. As he himself said, he had served without any ambition. *One was a soldier because Kaiser and fatherland wished it so, and one would have been embarrassed if one were to have been exempted from compulsory service.*

His supervisor Zincke let Hahn know during his military service that he would like to see him again in his institute on the 1 October 1902. With delight Dr. Hahn took the assistant's post offered and went back to the University of Marburg for two years. With Zincke's recommendation, so he hoped, he would find a good position in industry later.

Privy Councillor Zincke, who had held a chair in chemistry at the University of Marburg since 1875, and had admittedly provided no new ideas that had influenced the direction of chemistry as a science, had been one of those markedly productive in the field of systematic chemistry. As his most distinguished duty he studied the art of teaching. One of his pupils wrote of it in an obituary of his teacher who died in 1928, "Out of selfless dedication he daily imparted to everyone individually, from beginner to doctoral candidate, understanding of chemical action and the methodology of work in chemistry. With the greatest of order and tidiness the master educated his pupils in the indispensable prerequisites for successful creative work, the keenest observation and strictest self-criticism".

Doubtless Otto Hahn had much for which to thank his teacher Zincke, especially so his upbringing in systematic scientific work and exactness in working. For two years he assisted in Professor Zincke's chemistry lectures. In order to be fully prepared he had to arrive at the Institute shortly before eight o'clock, and his first job was to light the large gas flame under the lecture theatre's bench. A little later the Privy Councillor came, critically checked Hahn's preparations, and wrote chemical formulæ and equations on the blackboard. *Although I was not a very adroit experimenter*, recalled Hahn, *the lectures went quite well, and Professor Zincke was content.*

Zincke fired his students with enthusiasm through dramatic experiments. In the lecture about inorganic chemistry that took place in the summer half of the year the high point was the demonstration of exploding chlorine gas. For this purpose Otto Hahn had to place a two litre flask full of hydrogen and chlorine in a sunny spot, and everybody waited eagerly until the contents shortly thereafter reacted with a jet of flame and loud explosion. In another experiment Zincke's assistant filled soap bubbles with the usual explosive gas and brought them to explode with a candle. As a precaution the candle was made fast to a long pole. A great cheer went up every time in the lecture theatre if some of the nimbly floating soap bubbles sometimes got away and threatened the assistant Hahn. *Of course, they wanted the largest possible soap bubbles because they exploded especially loudly.*

In the lectures on organic chemistry there were fewer experiments. As a result Hahn found more time for other work. His dissertation, which, as usual, had already been published in 1901 as a monograph, he prepared jointly with Zincke as a new publication. This work appeared in a more wide ranging form on 18 September 1903, in the respected technical journal *Liebigs Analen der Chemie* under the title 'Mitteilungen aus dem Chemischen Institut zu Marburg'[1]. It was the first time that Otto Hahn's name was to be read in a scientific journal.

There were no further publications from his time at Marburg. As Hahn himself vouchsafed, at that time he had no ambition at all to be a researcher. This is quite remarkable because later he wrote more than two hundred and fifty scientific publications. In those days Hahn certainly did not have the benefit of the presence of a person from the life of science who would have awakened in him enthusiasm for research.

He did not wish to become a researcher, but much more to work in the area of organic chemistry in industry. That Otto Hahn solidly guaranteed. At that time organic chemistry was conducted according to a quite manageable set of facts; the chemistry of the benzines had just had its triumphant progress reported. Synthetic dyes began to win the hearts of the market and to break the monopoly of natural dyes such as indigo. Emil Fischer founded by his work the chemistry of carbohydrates and protein substances. Year on year new preparations enlarged the palette of synthetically manufactured medicinal substances. Professors of organic chemistry at universities were

---

[1] *'Communications from the Institute of Chemistry, Marburg'.*

unanimous, as were directors of institutes, that their field held the position of prime importance. Inorganic and physical chemistry had to wrestle for recognition and equal rights.

And so there was not a shadow of doubt for Otto Hahn that after the expiry of his time as an assistant his future was in working in organic chemistry. His plans seemed to him to be realisable, as he would receive the offer of a position as an industrial chemist through his doctoral supervisor. The firm, however, wanted Hahn to have a good command of foreign languages, because activities in other countries would be also amongst his duties.

Zincke advised Otto Hahn to go to England for a while and succeeded in obtaining for him, his protégé, a position with William Ramsay at University College, London. The two had been acquainted through shared time as students in Germany. Hahn's parents were able to finance their son for a stay abroad, and so did Otto Hahn journey to London in October 1904.

2.4 "You Will Work on Radioactivity"

At the beginning of the twentieth century the metropolis of the British world empire, with its more than five million inhabitants, was the largest city on Earth. London was the centre of world trade. Also the scientific life of England was concentrated in the capital. The Royal Society, in character like an Academy of Sciences, could look back on a two hundred and fifty year old tradition. In the annals of the society are found the names of the great British natural scientists, and in its Foreign Members are counted the most famous scholars of the entire world. The achievements of British science are embroidered with the names of Boyle, Newton, Dalton, Faraday, Maxwell, Darwin, Thomson, Ramsay, and Rutherford, who were especially important to the development of the natural sciences and helping The Royal Society gain its international reputation.

When Otto Hahn journeyed to London in 1904 to work at the famous University College in Gower Street, some seventy professors taught there for about sixteen hundred students. As a young scientist Hahn was very conscious of the honour that Ramsay's invitation meant for him. Sir William Ramsay was an outstanding scholar. He had been honoured with the Nobel Prize for Chemistry in 1904 for the discovery of the inert gases and the elucidation of the composition of atmospheric air.

Hahn told Professor Ramsay about his scientific career and requested a task to pursue. After a short consideration Ramsay said, "You will work on radioactivity".

For the organic chemist Hahn this bit of news came as rather a surprise. Of radioactivity he had heard not a single word in the lectures at the university. He was honest enough to own to the distinguished scholar that he understood nothing of radioactivity and that in this field he had not a shred of experience. Ramsay was a good psychologist. "That is exactly the right person we want. You have no preconceived ideas and therefore can be completely impartial in your approach for us in this very mysterious topic". And he then made the young chemist familiar with his well known research work in a spellbinding way.

The Englishman had purchased five hundred kilogrammes of rare thorianite. In the entire world it came only from the island of Ceylon (known today as Sri Lanka), and then was found only scantily. It was known of this mineral that it was strongly radioactive. A firm had already processed the rock at Ramsay's request. From the five hundred kilogrammes there remained eighteen grammes of a white salt—essentially barium carbonate—which produced an amount of radioactivity the same as that from nine milligrammes of radium. Ramsay's proposal was to separate out the expensive radium by Marie and Pierre Curie's fractionated crystallisation process and to convert a few organic salts in order to determine the molecular weight. In this manner he hoped to be able to fix the then unknown atomic weight of radium. This was the attempt that Hahn should undertake. From the eighteen grammes of the substance to extract nine milligrammes of a strange element seemed to the young researcher to be a very tricky undertaking. What is more, he lacked any kind of working hypothesis.

In order to acquaint himself with the unknown field of work and the special working techniques, Otto Hahn had to make a fundamental study of the literature, to make tools and test them, and to plan the calibration of the tests. In this respect he had some good fortune as William Ramsay had turned his interest to radioactivity for some time, so that various apparata and measuring devices were already at hand in the institute. Even so, Hahn needed a little time to become conversant with radioactivity and its still young history.

3

The Awakening
of the Natural Sciences

3.1 Puzzling Radiation

The discovery of the phenomenon of radioactivity led in a new period in the natural sciences which drew profound changes into the arenas of science and ideology. It began on that memorable day, 8 November 1895. The physicist Wilhelm Conrad Röntgen was experimenting in his Würzburg laboratory with cathode rays produced by electrical discharges in extremely low pressure gases. To his great surprise he detected that a single crystal of a fluorescent substance shone brightly, and happened to be well away from the gas discharge tube shrouded with black paper that stood on a laboratory bench. This crystal must have been flooded in invisible radiation. There was no other way of explaining the observed effect. As the researcher soon discovered, it was a matter of 'A New Kind of Rays'[1]—according to the title of his paper of 28 December 1895—which was secondary to the cathode rays. They possessed the remarkable characteristic of passing through a substance apparently unhindered. When he placed his hand on the tube, Röntgen could recognise its bones on a screen.

The puzzling x-rays, in 1896 already known as Röntgen radiation after their discoverer, caught the universal attention of the experts. Also the daily press continually had reports about the mysterious x-rays, which were credited with fantastic properties.

---

[1] *'Eine neue Art von Strahlen'*.

Looking back, the English physicist Ernest Rutherford said of the year 1895, "It marked the beginning of a new fruitful epoch in physics, with discoveries of seminal importance following upon each other in an almost unbroken sequence".

Researchers all over the world sought feverishly to solve the puzzle which these cathode rays and Röntgen radiation presented. Joseph J. Thomson of Cambridge University, one of the leading experts in the field of physics, was able to prove in 1897 that cathode rays consisted of an enormous number of small particles negatively charged. Later he introduced the name of *electron* for them. But the most surprising thing was that their apparent mass was almost two thousand times smaller than that of the lightest atom, the hydrogen atom. Up to then it had been believed that the atom was the smallest building block of matter.

There was an enthusiastic search for more invisible rays, which up to then might have escaped the gaze of scholars. In Paris the physicist Henri Becquerel experimented with uranium compounds. In these substances he wanted to test whether the fluorescence that emanated from uranium salts was also "a new kind of ray". An accidental discovery gave his investigations a completely new direction.

On 1 March 1896, the Frenchman was about to carry out his crucial experiment. To be on the safe side, however, he developed one of the photographic plates laid in a drawer together with the uranium salt. To Becquerel's astonishment, the topmost plate showed a distinct darkening, and, to be precise, of all places at the very position where the uranium compound had been. From where had this "incident light" come? In the drawer it had been pitch dark. There was only one explanation: the photographic plate must have been darkened by an unknown ray from the uranium compound. It was soon confirmed that the metal uranium also exhibited this 'radiographic effect'. Becquerel rays, also called uranium rays, revealed their ionising effect: the charged leaves of an electroscope rapidly fell closed if the air between them was ionised by these rays, that is to say, they were split up into electric particles. For these typical effects of the rays the term 'radioactivity' was coined. That 1 March 1896 should go down in history as the day in the annals of science of the discovery of radioactivity.

Uranium had already been discovered in 1789 by the German chemist and apothecary Klaproth, but was first obtained as a pure metal in 1842. Except for its occasional use for colouring glass and porcelain, and later in photography, no use for this element had been known for over a hundred years. Because of its rarity it decorated the shelves of chemical laboratories as a curiosity. Suddenly uranium had become interesting to science. The most important uranium mineral, pitchblende, would years later become one of the most sought after ores. Up to then, pitchblende was usually seen only by people in the mountains, for its occurrence indicated an impoverishment of a lode of silver ore. This black mineral had been thrown on the slag heap as useless.

Two years after Becquerel's discovery, in April 1898 his pupil Marie Curie-Skłodowska of the Académie des Sciences in Paris informed him that not only uranium but also the second heaviest element, thorium, emitted these puzzling rays, and thus was radioactive likewise. Madame Curie then made an even more significant discovery. Natural uranium bearing minerals like pitchblende were more strongly radioactive, as was to be expected on the grounds of its uranium content. Marie Curie's hypothesis, that contained in these minerals there must probably be an even stronger radioelement, found an outstanding confirmation. Together with her husband Pierre Curie, she succeeded in 1898 first in physically proving the existence of two new chemical elements. A year later she isolated from a tonne of residue, which resulted from the use of the Joachimsthal pitchblende, a few milligrammes of salts of these basic chemical substances. These two surpassed uranium in their intensity of radiation by many times. The couple Curie called them polonium in honour of the homeland of the lady researcher, and radium. A year later the French chemist André Debierne recovered a further radioactive element from pitchblende residue, actinium.

For their pioneering achievements in the field of research into radioactivity Becquerel and the Curies were to be awarded the Nobel Prize for Physics in 1903. The first Nobel Prize for Physics Röntgen had received in 1901.

Today it is only with difficulty that one can imagine under what primitive conditions the researcher of radium made her fundamental discoveries. An old shed meagrely provided with equipment served Pierre and Marie Curie as a laboratory. The most important instrument for their work was a simple electroscope with two thin gold leaves. With this 'bloodhound' the Curies pursued step by step the enrichment of the unknown radioelements in the fractionated substances.

In the early days of exploration of radioactivity, many scientists made this single but important piece of equipment themselves. Ramsay used an empty tin oil can in which he cut two window-like openings so that it was possible to observe the gold leaves fixed to a wire in the middle. Despite its simplicity this gold leaf electroscope "proved to be a precise and reliable measuring instrument, and played an important role in the exploration of radioactivity". Ernest Rutherford, one of the great pioneers, delivered this assessment in 1906. He was a superb master of improvisation if physical measurements were to be made with the simplest aids.

Experimental scientists of that time had to have at their disposal solid skill in handiwork. Physics equipment as good as was needed was not available from business. The physicist Lenard himself singlehandedly wound the induction coil needed as a high voltage generator, and needed several weeks to do so. A scientific periodical wrote with meaning, in 1898, "The owner of a good induction coil looks after it like a virtuoso violinist cares for his

Stradivarius". William Ramsay built his numerous apparata for testing for gases himself. He was an excellent glass blower.

It is understandable if a few of the great experimental geniuses of that epoch distrusted the complicated instrument parks of the modern age. So once thought the English physicist Lanchester unimpressed by the inspection of a modern research laboratory, "Much too many pieces of equipment, far too few brains".

3.2   Pioneering Ideas

Radium began to fascinate the world. Its radiation effects were many times stronger than those of the other radioactive substances. The newly discovered element seemed to be a spring of inexhaustible energy. In addition no diminution in the intensity of radiation was observed over a long time. Today we know that the former falls by a half after 1,590 years.

In the very first years after the clear description of radium there was such an abundance of reports communicated about the increase, decrease, and also the change of character of the radioactive effects, that everything seemed to be an unravellable chaos, inaccessible to the methods of exact investigation. Thus Otto Hahn described it in 1907 in considering the situation in which he found himself to be in the exploration of radioactivity at the start of the century.

It is to his lasting credit that the English physicist Rutherford brought light into the darkness of mutually contradictory observations and cleared up the mystery that was radioactivity. Born in New Zealand, Ernest Rutherford had studied at the famous Cavendish Laboratory at Cambridge under Professor Joseph Thomson. At the age of twenty seven, he went to Montreal in 1898 to undertake the direction of his own research institute.

"I have never had a student with more enthusiasm or ability for independent scientific research than Mr. Rutherford, and I am sure that in Montreal he will create a respected department of physics", sang the testimonial of his teacher Thomson.

In Montreal Rutherford devoted himself, with a manifest intent for the future significance of this area of research, exclusively to radioactivity. In a short time he succeeded in making scientific discoveries that changed the conception of the physical world. First Rutherford showed that there are three different kinds of radioactive rays, which he named alpha, beta, and gamma rays. Beta rays—which were discovered first—exist as the same negatively charged elementary particles (electrons) as cathode rays. Their speed amounted to over 200,000 kilometres per second! Alpha particles possessed a larger mass, and also were catapulted at high speed—around 15,000 to 20,000 kilometres per second—from out of the radium atom.

In the search for a comparison, alpha particles emitted from an atom were compared to shotgun pellets, the enormous penetrating power of which is known. But how inadequate was such a comparison. A rifle bullet leaves the barrel with a speed of one kilometre per second; against that an alpha particle leaves a radium atom with a speed twenty thousand fold. *With such an initial speed it succeeded in out-firing a shell*, explained Otto Hahn in a talk about the enormous energy of alpha rays, *then, if one wished, one could fire to the Moon, for a shell with such a speed would easily overcome the Earth's attraction.* Owing to their minuteness, alpha particles, however, travel only a small distance. But despite a small range of only a few centimetres in air, hundreds of thousands of gas molecules are ionised by a single alpha particle. Such a bombardment as over thirty six million alpha particles emitted per second by a milligramme of radium at the microscopic scale appeared to be inconceivable, so it was thought.

The conclusive elucidation of the nature of these rays took over ten years, until it was demonstrated that alpha particles are the nuclei of helium atoms and gamma rays a kind of Röntgen radiation.

To Rutherford's credit also belongs the explanation in 1900 of so called radioactive emanations. He confirmed the reports made by other investigators in which the radioactive elements thorium, radium, and actinium discharged gaseous products, called emanations, which are likewise radioactive and disintegrate in a short time.

Rutherford's collaboration with the chemist Frederick Soddy, who had worked as his assistant since May 1900, proved to be fruitful. Soddy was, like Rutherford, an experienced experimental scientist. Their joint publications 'The Cause and Nature of Radioactivity' (1902) and 'Radioactive Transformations' (1903) made Rutherford and Soddy known for a theory of radioactive decay which proved to be seminal for further research. They perceived radioactivity as "a consequence of a process" that lay entirely outside the sphere of all known forces, and which was not created, and "could be neither changed nor destroyed".

According to this theory of disintegration, the atoms of a radioactive element are unstable, and as a consequence have only a certain characteristic life time. This introduced the concept of 'half life', during which half of the number of atoms in a radioactive substance would decay. Through this process radioactive elements changed into a series of other substances which were no longer identical to the mother element in their chemical nature. The radioactive decay, so claimed Rutherford and Soddy, is not dependent upon external conditions. Completely new dimensions would have to be come to terms with in the interpretation or radioactivity. It was ascertained that in a gramme of uranium ten thousand radioactive atoms decayed, and in a gramme of radium more than thirty million atoms. Nevertheless, these numbers are modest in comparison with the total number of available atoms. One gramme of radium contains more than ten billion billion atoms, 2.66×10^{21} to be exact. The proportion that change in one second is therefore so tiny that thousands of years would have to pass before radium is completely decayed and converted into lead.

The publication of the theory of radioactive decay caused a sensation. There was enthusiastic approval, but also passionate rejection. Rutherford's book *Radio-Activity*, which appeared in 1904, was immediately sold out, which happens to scientific publications only rarely. Fantastic things about radioactivity were also to be read in the press. Journalists, ever hungry to find out news, utterly laid siege to the few laboratories there were at that time carrying out radium research.

Rutherford's name became known over all the world. Invitations to lectures and countless honours were heaped upon him. The physicist from Montreal was regarded as the uncrowned king of the field of radioactivity. His teacher Thomson could not conceal his admiration when, in 1904, he wrote of him, "Rutherford has not just widened the boundaries of our knowledge, but has taken a new province by storm".

3.3   'Atomic Energy'

But what really racked the brains of the academics was the apparently inexhaustible energy of radioactive elements. Their unchanging strong rays, their light in the dark, the higher temperature of radium salt solutions than their surroundings—everything that was appropriate to prompting fantasy.

At the beginning of October 1903 Rutherford gave a lecture in St. Louis "On Radium" and attempted to describe what unbelievable amounts of energy were in play in radioactive processes: one pound of radium emanation— it was assumed that one could gain access to this energy—shines out energy incessantly, which corresponds to a performance duration of around 10,000 horse power. This amount is millions of times greater than obtainable by chemical reactions, for example, in the burning of coal.

Whoever did not hear Rutherford's lecture read about it in the press. For many visitors from outside the city who went to see the preparations for the 1904 World's Fair, the *St. Louis Post-Dispatch* of 4 October 1903 reported about the new source of energy. The newspaper was already speculating about its use in a war which might destroy the entire Earth. For the first time words were used which henceforth were bound to hope and fear. One phrase held in itself overtones of the inconceivable and the monstrous— *atomic energy*.

"The emanation is a source of gigantic energy", enthused Ramsay on the occasion of a lecture tour in Vienna in April 1908. "A cubic centimetre, if we could collect that much, would through its decay give more heat than three cubic metres of exploding oxy-hydrogen ... In a word, in emanations one has a chemical weapon which outdoes the power of the usual reagents in the same way that a modern rifle does the bow of our ancestors".

Ramsay's example had only a theoretical interest, and never came to any practical significance. But not a few would have looked forward with fear to that day in which mankind would succeed in reality in setting free the energy of the atom.

But the popular lectures and presentation of a few atomic researchers did not lead to a spread of such fears. Rutherford held his lectures "On Radium" also for a circle of scientists. On 30 December 1903 he went before the American Association for the Advancement of Science, "If a suitable detonator were to be found it can be imagined an explosive wave of atomic decay could propagate through the material which could reduce this ancient Earth to ashes". Rutherford did not hold his fears back, "Any simpleton, unseen in a laboratory, could blast the entire world into thin air".

Pierre Curie also was beset with similar such reservations because of his conviction that "It would be very dangerous for radium to fall into criminal hands". In his speech given in June 1905 on the occasion of the award of the Nobel Prize he raised the question of "whether mankind gains profit from knowing the secrets of nature, and whether mankind is mature enough to make use of them, or whether this knowledge is harmful".

Frederick Soddy had occupied himself with the problem of radioactivity as deeply as Rutherford. His statements are astonishing for their foresight of the future. In January 1904, Soddy gave a lecture before the military engineering school of the English naval port of Chatham, and again a few weeks later in Manchester before a literary–philosophical society in the city. In front of this most varied circle of listeners, Soddy developed his thoughts about atomic energy: "If it could become accessible and controlled—what a tool it would be, to be at the controls of fate of the world! Man would possess a weapon with which he could destroy the Earth if he wanted to". Soddy did not conceal from the military the danger he had seen. He even exhorted his listeners "to hope with him that Nature would keep its secrets".

But Soddy himself, who early recognised the conflict in which atomic research could be caught, showed himself to be fascinated by the thought of releasing this fantastic energy. In his book *Radio-Activity*, which appeared in 1904 shortly after the identically named publication by Rutherford, he pointed to a way in which this inexhaustible source of energy could be managed: "It is known that such radioactive elements as radium and uranium convert into radiation energy over a period of thousands, even millions, of years by emission. Soddy astutely concluded that all this energy would then be available to all men in the future "if this conversion time were sped up and the very considerable amount of energy that is now spread over a thousand years could be extracted for the immediate use" ".

The Berlin chemist Willy Marckwald had similar thoughts when he addressed the meeting of the German Chemical Society[2] on 2 May 1908 about the miracle of radioactivity. "We know of no method of speeding up radio-

---

[2] *'Deutsche Chemische Gesellschaft'*.

active decay", admitted the lecturer. "If we had such, with its help presumably we would be capable of transforming other elements. Thus we would have to expect the formation of elements with low atomic weight with a simultaneous gain of an enormous quantity of energy. If the conversion were suddenly to take place it would have to be accompanied by a most terrifying explosive effect. If it were, however, controllable, a kilogramme of pitchblende would certainly be sufficient to transport a large, fast steamer across the Atlantic Ocean".

In spite of such fears investigations were being made into whether the fundamental problem of the decay of radioactive substances was to a certain extent a catalytic speeding up to free the atomic energy spontaneously. Rutherford set the temperature of radium emanation as 2,500 degrees centigrade and a pressure of 1,000 atmospheres, which would be sufficient to explode a steel casing. But the radioactive characteristic decay of the emanation remains the same.

In the mean time it was then sought, but still through 'classical' experiments, to influence the steady decay of a radioactive substance. But the half life did not change, either when subjected to extreme cold near absolute zero or to extreme heat, whether under a pressure of two thousand atmospheres, whether subjected to a magnetic field of up to 83,000 Gauss, whether in deep mine galleries in the Jungfraujoch, or when subjected to centrifugal forces thousands of times stronger than the attraction of the Earth.

Nature did not allow itself to be outwitted by such methods.

3.4   Radium—The Great Revolutionary

Inevitably the decay theory of Rutherford and Soddy had to lead to new epistemological conclusions, for they shook the classical ideas of natural philosophy and their seemingly so firmly established theoretical foundation.

Up to the end of the previous century the conception of the world worked out by Newton from its foundations had been flawless. Within it space and time were absolute concepts and all physical events behaved according to the fundamental laws of mechanics. The world was built not of divisible particles; namely, not out of atoms and elements constructed out of the latter. 'Atom' was taken from the Greek word 'atomos', by which was meant nothing less than 'indivisible'. In that context one sought to demonstrate that atoms and elements were unified and impenetrable and would not be able to change from one into the other.

With the discovery of x-rays by Röntgen, the concept of the impenetrability of matter had become invalid. Matter was no longer a hindrance to these mysterious rays. When Thomson proved the existence of electrons the proposition was invalidated that the atoms were the smallest building blocks of matter. Rutherford's researches thus brought further evidence.

The spontaneity of the radioactive decay and the transmutation into another element appeared to be inexplicable. A further dogma was violated in addition. Alchemists had for centuries dreamed of the transmutation of an element, and now it appeared to have been demonstrated, at least at the microscopic level.

In 1905 the French physicist[3] Henri Poincaré alarmed the experts with his doubt about the 'value of science'. That was the name he gave to his paper which he published in 'Signs of a Serious Crisis in Physics'. The 'great revolutionary, radium', not only brought the principle of the conservation of energy into question, but also all the other maxims of science. "Before us lie the ruins of the old principles of physics, the collapse of which we are experiencing", lamented Poincaré. Also the previous scientific concept of matter had become untenable as a result of radioactivity. New conceptions about the structure of matter and their manifestation began to determine the character of the view of natural science. A few pioneering achievements began to take on a special role. Such was the discovery of the elementary quantum of energy by Max Planck in 1900, which was an epistemological turning point. Albert Einstein proved the quantum nature of light in 1905, and blew apart the dogma of the previous mechanistic ideas about the absolute nature of space and time with his relativity theory. With his famous formula $E = mc^2$ he provided the long sought relationship between mass and energy.

Up to that time all estimates of the energy of atoms had been speculative, and the predictions of its technical exploitability had been, to say the least, premature. But now the conversion of atomic mass into energy had become theoretically grounded. The idea of using atomic energy in a practical way for the first time had obtained a solid foundation. By Einstein's equation— so the scientists soon calculated, to the astonishment of the public—one gramme of matter could, by its complete conversion into energy, release the equivalent of twenty five million kilowatt hours. This amount corresponds to the heat obtained by burning two hundred and fifty wagons of coal!

In the search after an answer to the question of how this conversion of matter into energy could be brought about, one was again always led to the example of the radioactively decaying atom. Soddy vividly compared the spraying of energy by the radium atom with Aladdin's Magic Lamp from the *Tales of a Thousand and One Nights*. This graphic comparison Soddy used in a series of lectures which he gave in Glasgow in 1908, and in the following year under the title of 'The Interpretation of Radium'. We are compelled to respect Soddy's forecast that the source of energy of uranium is 'more miraculous' than that of radium. If only a way could be found to speed up the slow decay of uranium over a million years. Such a process, of course, would—so thought the co-founder of the theory of

---

[3]See Note 7 in Translator's Notes on the Text.

atomic decay—only be found "if we can bring about the conversion of an element at will"— a quite astonishing statement, thirty years before the time when a solution of the problem actually became available.

But Soddy left no doubt in his observations of 1908 that "a day will come in which we will be able to split the elements in the laboratory and rebuild them". Energy would then be available in abundance. "If mankind were able to convert the elements they would not need to earn their bread by the sweat of the brow. We can easily imagine that such a people will make barren continents fruitful, bring ice from the pole to melters, and make it possible to turn the whole globe into a paradise".

However, this paradise on Earth invoked was not the only path that mankind would open up for itself with the help of atomic energy.

4

The First Scientific Discovery

4.1   In London with William Ramsay

With the announcement that gaseous radioactive emanations behaved like
inert gases Ramsay's interest in radioactivity was awoken. The discoverer
of the rare gases argon, helium, krypton, and xenon was aroused to solve
yet another puzzle. It had been found out that helium occurred in all radioactive minerals. A convincing explanation was lacking for each of them.

In collaboration with Soddy, who had returned to England in 1903,
Ramsay attempted to find experimental proof that helium emerged from
radium. Rutherford had already voiced this supposition in 1902 in the interpretation of radioactive decay.

At the beginning of 1903 it was first possible to obtain small quantities of the rare radium commercially. In the entire world there was only
one single source of supply, Professor Friedrich Giesel in Brunswick, who
carried on the manufacture as, so to say, a hobby. Ramsay and Soddy obtained a quantity of thirty milligrammes and immediately set off on their
experiments.

In actuality the pair succeeded in substantiating the formation of helium
out of radium with the help of the sensitive methods of spectral analysis.
Owing to the tiny amounts of substance the investigation was very intricate.
All of Ramsay's self-designed apparata had to be built to the most exacting
standards—out of little capillary tubes of less than half a millimetre in
diameter. Ramsay and Soddy published the results of their 'Experiments
on Radium and the Production of Helium from Radium' in July 1903.

For the first time it had successfully been proved experimentally that one chemical element had converted into another.

When Otto Hahn was given the task by Ramsay of working in, for him, the new area of radioactivity he set to the work with zest. He studied the available literature and attended Ramsay's lectures. Amongst other things, the English academic presented a detailed paper in his talk, given at The Royal Society on 28 April 1904, on the theme 'Further Investigations of the Production of Helium from Radium'.

Hahn carefully noted down what Ramsay reported in his talk. These minutes have survived. They afford us a comparison with the text and the contents of Ramsay's published works. Sketches and attempts at descriptions prove that Hahn reported accurately and was conscientious in his records. These characteristics served him later, also, as a successful researcher of the highest calibre.

After the visit to the lectures Hahn devoted himself exclusively to the solution of the task given to him by Ramsay, the separation of nine milligrammes of radium from the residues of Singhalese minerals. In doing so he had some surprises, which he at first ascribed to his inexperience in work in radiochemistry. These lay in finding that the end products were still always radiating radioactively, although he had already separated the radium. He repeated the operations and set about a still more careful separation. But a different result did he not receive.

In the young radiochemist there awoke the thirst for knowledge. What substance could have been left behind in the product? It was not radium, and the other known radioelements did not exhibit such strong radioactivity. To his astonishment Hahn established in addition that the unknown carrier of this radioactivity did not display the characteristic emanation of radium but of thorium. From that one could conclude that the product contained thorium. But this element is not nearly so radioactive.

The emanations of radium and thorium can be distinguished well by their decay times. To do so Otto Hahn used an experimental set-up of Rutherford and identified the emanation either with the help of a fluorescent screen or an electroscope. In his portrayal in the *Jahrbuch der Radioaktivität und Elektronik* of 1905 he described his experiment as follows: *In the form of bubbles through a fairly long pipe a stream of air was sent through the end product, this being performed in a darkened room; after the eyes had become accustomed to the darkness they were directed to the zinc sulphide screen, where shortly thereafter it was seen to glow* . . . .

4.2   Radiothorium—The First New Element

Finally the young researcher succeeded in separating a tiny amount of the unknown substance from the radioactive end product. It was not even half a milligramme that would be *tested for activity in the electroscope*. The effect was amazing. Hahn measured an extraordinarily strong radioactivity. *The electroscope closed up in something like two seconds, whilst by comparison five milligrammes of uranium oxide took seven and a half minutes to do the same thing.* The new radioelement could not be chemically distinguished from thorium, but was, however, much more strongly radioactive. Because of this Hahn called it radio-thorium for short.

It was a pure chance discovery, he later said, honestly. Without wishing to diminish the credit of the young scientist we must add that in reality there was a bit of luck in his first discovery. A colleague, Gian A. Blanc from the University of Rome, had likewise been hard on its track. Blanc had made for himself the study of radioactive sources from Echaillon and Salins Moutiers in the Haute Savoi region of the Alps. In the deposits of these substances he found a radioactive product similar to thorium 'unbelievably more radioactive than thorium salts', recognisable by the short lived thorium emanation.

Already in January 1905, chronologically before Hahn, Blanc had published a paper about it in the *Philosophical Magazine* in which he announced further attempts. For some reason he delayed it. When Blanc wanted to announce the results he had found, he must have found out that Ramsay had already given a report in Liège, in September 1905, about the radio-thorium obtained by Hahn from thorianite. Controlled experiments showed Blanc that the product he had isolated corresponded to Hahn's radio-thorium.

Ramsay was very enthused when yet another new element was found in his institute. He intended to announce the discovery in a correspondingly suitable way. In accordance with tradition this should be done before the committee of the venerable Royal Society.

At the session of The Royal Society on the 16 March 1905 Ramsay communicated the discovery of radio-thorium. For the first time the name of Otto Hahn was mentioned in connection with radium research. The publishing medium of the Society *Proceedings of The Royal Society* contained in the issue of 24 March 1905 a paper entitled 'A New Radio-Active Element, which Evolves Thorium Emanation. By O. Hahn (Communicated by Sir William Ramsay)'. It was the first of more than two hundred and fifty scientific publications by Otto Hahn in the field of radiochemistry.

The German technical journals also published the article; the first was the *Zeitschrift für physikalische Chemie* in its issue of 9 May 1905, as 'Ein neues radioaktives Element, welches Thoriumstrahlung aussendet'. In detailed form, with all the experimental particulars, Hahn's contribution appeared in the *Jahrbuch der Radioaktivität und Elektronik* 1905, from which we have already cited a few experimental details.

Even the daily press was interested. The *Daily Telegraph* announced on 8 March 1905, Hahn's birthday, "a new discovery which has been added to the many brilliant triumphs of Gower Street. Dr. Otto Hahn, who is working at University College, has discovered a new radio-active element, extracted from a mineral from Ceylon, named thorianite... ".

A letter from Hahn to his student friend George Dahmer has been found, dated 24 March 1905, which tells us a few details of Otto Hahn's work in Ramsay's institute. Amongst other things we read, *I am now also trying my hand in the area of inorganic, to be precise, physical, chemistry; as you know, I have also obtained a quite beautiful result which Sir William Ramsay recently presented at The Royal Society.... My customary good luck has held and not deserted me here in London. Incidentally, I have not been idle. For many weeks I have been going into the laboratory again, with few exceptions, at eleven or half past to make brief measurements, during the day itself they are made inaccurate by disturbances from the local commerce and industry. Ramsay gave me a key which opens all the doors in University College so that I can get into the physics and chemistry laboratories at any time....*

4.3   An Uncertain Future

Ramsay had an open ear for the personal problems of his students. When the end of Hahn's stay approached, Ramsay asked him about his plans for the future. Hahn told of his employment as an industrial chemist. Ramsay shook his head disapprovingly.

"Mr. Hahn, you have discovered a new radioactive element. That is an excellent recommendation. For that reason you should remain in radioactivity. Go to Berlin. A place is certain to be given to you in the greatest German university, where you can carry on your researches and qualify as a university lecturer... But before that you should work under Rutherford in Montreal. Nowhere can you learn more about radioactivity than in his institute".

On the same day, which was 26 March 1905, Ramsay wrote to Emil Fischer, the Director of the Chemistry Institute at Berlin University. Ramsay was friends with Fischer, who had repeatedly invited him to Germany to give talks, and who had made him very much at home in his hospitable house.

After Fischer's letter of agreement was received in London, Hahn applied to Rutherford. Ramsay added a few words to the letter of recommendation of 20 May 1905: "Hahn is a splendid colleague and has done admirable work. I am sure you will enjoy working together with him".

Ramsay's letter to Fischer has been preserved. In impeccable German Sir William reported the history of the discovery of radio-thorium, giving his protégé a good report.

"I have been quite astounded by the boldness, skill, and stamina of Dr. Hahn. Naturally, I have spoken to him daily about his problems.... Hahn has also studied under Zincke at Marburg. He should qualify as a lecturer, and I believe it would be good if he did so with you. Would it be possible for him to work in your laboratory for a couple of years? He is a nice fellow, informed, trustworthy, and talented, and I have come to like him very much. He is, and wishes to remain, German, and may be trusted with all methods of investigating radioactivity. I know that you want to make your laboratory as versatile as possible; do you have a corner for him? ... I can recommend him as one of the best workers whom I know".

During the summer of 1905 Hahn visited the famous chemist and Nobel Prize winner in Berlin. For the time being Emil Fischer indicated that there was no particular department in view in which he could put him, but could only offer his help if Hahn wanted to qualify later as a lecturer.

Despite Fischer's cooperation Otto Hahn's future thus remained largely uncertain. The university path of a radiochemist was at that time extremely unfavourable. The chairs in German universities were occupied by professors of organic and inorganic chemistry. There were no lecture courses about 'radioactivity', and there was as good as no research. Otto Hahn had second thoughts. When he was considering what was best to do Rutherford's invitation arrived. He was pleased to receive it and and in his letter of acceptance to Rutherford he mentioned that he had discovered a new radioactive element.

But this communication was received in Montreal with some reserve, as Otto Hahn was to learn later. A new radioelement? From a thorium rock? A few years before—in 1901—the American Charles Baskerville had found the alleged 'new' element Carolinium in the thorium bearing monazite sand of North and South Carolina. It was an error. To his doubt, Rutherford saw in a reference that it was confirmed by his friend Bertram B. Boltwood, who taught as Professor of Radiochemistry at Yale University. Boltwood wrote without beating around the bush, on 22 September 1905, of Hahn's 'discovery' to Rutherford, that "The substance appears to be a new compound of thorium X and stupidity".

The daily papers and the scientific journals on many occasions reported the discovery of a 'new' element. They shot into prominence and disappeared like comets, being called Nipponium, Berzelium, Carolinium, and so on. Ramsay himself mistook one such sighting when he reported before The Royal Society that a Japanese scientist at his institute had discovered the new element Nipponium. Ramsay had to correct himself later.

4.4   The Finest Year of His Life: In Montreal with Rutherford

Divided feelings beset the young scientist when he set sail for New York in the middle of September 1905. Would he survive under Ernest Rutherford as well as he had under William Ramsay. The silhouette of New York, which began to emerge over the horizon, brought other thoughts to Otto Hahn.

With its high rise buildings, the city presented an impressive sight which was completely different from the appearance of european cities.

After a few days' stay in New York Hahn left the city and went up the Hudson River by steamer to Albany. From there he travelled on to Buffalo and to Niagara Falls. He finally arrived in Montreal by rail.

The bustling port and industrial city numbered some two hundred and seventy thousand inhabitants. Montreal was the largest city of the British Dominion of Canada. It accommodated English and French Canadian communities. The boundaries were invisible. Hahn, as always, prepared himself for new pleasures, even if the guard of the electric train greeted his passengers with "Tickets, please, gentlemen" and a few stations later called out "Les billets, messieurs".

Two universities gave the scientific face of the city its character, the catholic Laval University in the French quarter and McGill University in the city's English quarter, in which was Rutherford's work place in the MacDonald physics building.

Rutherford immediately caught up his guest in a long conversation about the allegedly newly discovered radio-thorium. Rutherford did not rein back his doubts. But Hahn was able to allay Rutherford's scepticism, especially that he did not rely on his luck, for shortly thereafter he was to find in that institute a further unknown radioelement that had been overlooked there, radio-actinium.

In order to be able to work undisturbed Ernest Rutherford had put his measuring apparatus in the cellar of the building. Hahn's work place was also there. Rutherford's research was entirely into his favourite topic, alpha rays. For his guest, he had a similar task ready. Hahn was to investigate the nature of alpha rays that he had discovered to be emitted by radio-thorium. To this end he identified the range of the alpha particles of radio-thorium.

The favourite was the scintillation method; the particles were made to collide with a fluorescent screen, and by suitable magnification one could perceive shining points of light which flashed in a constantly changing way. "The phenomenon was an extraordinarily beautiful sight", wrote Rutherford in his textbook *Radio-Activity*.

As Otto Hahn used these methods he gathered the results in a manuscript which he completed on 26 March 1906. We can gather evidence from the text of his publication 'On Some Characteristics of the Alpha Rays of Radio-Thorium', which appeared in two issues of the *Philosophical Magazine* in June and July of 1906, and at the same time in *Physikalischen Zeitschrift*.

A small zinc sulphide screen on glass was placed vertically under the activated wire. The wire was placed on a small table which could be set at any desired distance from the screen by a to and fro motion in the vertical direction. As long as the screen was within the ionisation region of the alpha

rays it was possible to see the scintillation without difficulty with the help
of a magnifying glass. Above all it was necessary for the eyes to become
accustomed to the darkness....

Hahn had composed the English language text without the help of Rutherford. The publication was seen by his mentor during a lecture tour across the United States, and to Hahn he wrote on 20 August 1906, "In California I saw that your two papers have appeared in *Philosophical Magazine*. Both read very well".

Hahn's pleasure at the progress of his scientific work was also reflected in his eagerness to publish. A further manuscript about 'A New Product of Actinium', which was called radio-actinium, he completed on 6 April 1906. He submitted it to the well known London periodical *Nature* and to the publishers of the *Berichte der Deutschen Chemischen Gessellschaft*. In May 1906 the two illustrious journals published Hahn's work.

Under the instruction of Rutherford, who had developed a special apparatus for the purpose, Hahn decided on another experimental direction, the charge to mass ratio of alpha particles. The two found out that these particles always have the same mass, independently of from which radioelement they came. Rutherford and Hahn published their results jointly in issue number seventy of the *Philosophical Magazine* in October 1906.

His luck as a discoverer also did not fail him in Montreal. Apart from radio-actinium, in Rutherford's institute Hahn found a further decay product, thorium C', so that the former finally thought, shaking his head, "Hahn has a special nose for discovering new elements".

Even in the MacDonald physics building, a private foundation, one had to busy oneself with all the simplest physical apparata. Electroscopes for measuring alpha and beta rays Rutherford's coworker made himself from empty jam jars and tobacco tins. For the investigation of alpha rays that apparatus was to be pumped free of air, which was done with an antique pump of poor performance. Everything needed his time. So it was not seldom that the laboriously made active preparations had already decayed again before the measurements could be begun. But at that time—as Otto Hahn liked to emphasise—*with primitive means one can easily experience the joys of discovery.*

The atmosphere in Rutherford's institute was unconventional and friendly. The liberal deep discussions about only factual arguments Hahn felt stimulating. With many of his colleagues at McGill University he maintained a fruitful exchange of views all his life. Boltwood and Hahn became good friends after they were able to settle a controversy about the half life of radio-thorium. This difference of opinion arose because both researchers had found differing values.

Boltwood wrote of this to Rutherford, "Hahn must have worked incorrectly".

Hahn countered, "Boltwood must have made wrong measurements".

At Rutherford's suggestion, the matter was brought to a personal discussion. Both adversaries involved themselves in discussions, and each maintained to the end that he was right. At last Hahn realised that "We can both be right if we use as a basis the hypothesis that there must exist a transitional substance between thorium and radio-thorium which we have not yet discovered". A short time later in Berlin Hahn found the unknown link in the thorium series. He called it mesothorium. Boltwood was the first to congratulate him.

Ernest Rutherford was a fascinating person. Once he was possessed by an idea nothing stopped him from talking about it all the time to everybody. Thus it could happen, as Otto Hahn used to tell, that in colloquia which might be devoted to quite other topics he would report his newest investigations into his beloved alpha rays. Everything else was then forgotten. Any protest was silenced simply by his authority.

Hahn thought of the person of Rutherford always with genuine admiration, for *despite all his great successes Rutherford always remained the simple, happy, warm hearted person he had been from time immemorial, and all his students were devoted to him almost with a tender love*. Rutherford's bubbling enthusiasm for scientific research and his joy at work were proverbial, and awoke in Otto Hahn the love of science and a restless spirit of research. In proceeding with the elucidation of a scientific problem Rutherford was such an exemplary model of determination and tenacity that Otto Hahn made this mode of working his own. Doubtless this explains many of his later scientific successes.

Rutherford and Hahn parted as friends when the German left Montreal after three quarters of a year and went back to Berlin. The two maintained a lifelong animated exchange of ideas, as countless letters bear witness. The months in Montreal remained for Otto Hahn into his late old age the *finest time of my life*.

5

Research at the University of Berlin

5.1 The Joy of a Discoverer in a Workshop

When the new Chemistry Institute of Berlin University was opened in 1900, the members of the faculty declared that it ought to be the best equipped place of research in the world. In Emil Fischer the Institute had an internationally established scientist as Director.

The foundation of this institute was doubtless an expression of the growing significance which chemistry, especially in Germany, had gained in the world during the last third of a century. In comparison with England and France the German Reich had colonial possessions which were relatively small and poor in raw materials. In order to remain competitive in the world market important products would have to be produced within its own borders. That led to a strong build up of fundamental and applied research. Examples of this scientific race were the synthesis of artificial dyes, fuels, and artificial rubber, the production of high quality chemical products, especially of medicines, and the development of technological establishments for large scale synthesis. As a lucrative branch of industry the chemical industry brought the German capital market high profits and thus became a favoured object of investment. Powerful chemical groups arose which expressed their influence in the research establishments of the universities and high schools.

For the high standing of the science of chemistry in Germany spoke also the fact that half of all the Nobel prizes for chemistry in the first ten years in which they were awarded—1901 to 1910—were given to German researchers. Emil Fischer was the first German to be the recipient of the Nobel Prize for Chemistry, in 1902, for his pioneering work on the synthesis of sugar. He was the leading experimental chemist in the field of organic synthesis, and worked closely with the chemicals industry.

Although he was a specialist in the field of organic chemistry, Emil Fischer maintained a perspective of the entire branch of the science. He was concerned that applied and general chemistry be supported in his institute as well. At the start of October 1906 Fischer also made 'a corner free', an unused wooden workshop on the ground floor.

But it made little impact on Hahn. He promptly exchanged the old carpenter's bench for a long oak bench so that he could set out his measuring apparatus in order. From the Institute's workshop he got an electroscope to make his measurements, not one made out of empty jam jars and tobacco tins but of a fine brass case. Instead of the not very effective sulphur he used amber for the insulation of the small metal leaves. The charging of the electroscope was carried out according to the classic ritual in which one took a hard rubber bar and rubbed it vigorously on the sleeve of one's jacket.

First I naturally had a light fever. How would my future develop? Hahn's thoughts seemed understandable. In Fischer's Institute no one knew precisely what radioactivity was, and hardly anyone took Hahn's work in earnest. The young radiochemist was alone and thrown back on his own resources.

Otto Hahn had gotten radio-thorium as a 'dowry' from Ramsay. He bought two milligrammes of scarce radium from Giesel in Brunswick at the acceptable price of two hundred marks. Later the price of radium shot up when medical doctors' interest in this miraculous substance was awoken—radium could cure a cancerous ulcer.

When Otto Hahn also was in possession of the other important radioelements he would be able to continue his work begun with Rutherford. And his series of successes did not come to an end. In old thorium preparations Hahn tracked down the long sought decay product, the existence of which he had suspected in Montreal. Hahn called the new radioelement, which is very similar to radium, mesothorium.

Hahn completed his first manuscript about 'A New Intermediate Product in Thorium' in March 1907. On 13 April the *Berichte der Deutschen Chemischen Gesellschaft* published his new discovery. Parallel to it an identically worded article appeared in the *Physikalischen Zeitschrift*.

Hahn's mesothorium was to obtain significance as a less expensive replacement for the noticeably increasingly expensive radium. For a time it gained a high price as 'German radium'. The manufacturer of the radium replacement was at first the Berlin firm of Knöfler & Co. Later the Auer Gesellschaft produced mesothorium once they had succeeded in obtaining a patent for their process.

After radium, mesothorium was the second radioelement which would be manufactured in large quantities in moderate sized factories. As production material imported monazite sand was used, the basic raw material for thorium. For his collaboration Otto Hahn occasionally received from the firm of Knöfler very considerable remuneration, totalling over one hundred thousand marks. World war and inflation soon melted away this credit balance, which Hahn eventually exchanged for just one dollar.

Further investigations showed that this substance is complex, and is composed of mesothorium 1 and mesothorium 2. Hahn's endeavours led to the name of 'berlinothorium' for mesothorium 2. The laboratory assistant in Knöfler's thorium factory found a permanent name. She called it 'sunshine' because in the dark the enriched salts of this radioelement shone quite beautifully.

When Hahn's mesothorium seized further fields of application in medicine the discoverer drew up detailed information which appeared in the *Chemiker-Zeitung* on 3 August 1911, entitled 'On the Properties of Technically Manufactured Mesothorium and its Dosages'[1]. Interested persons learnt, to their surprise, that the new preparation was not actually an absolutely complete replacement for radium. Hahn conceded that in mesothorium there was always a contamination of twenty five percent of radium.

The faculty members were surprised, having come to value Hahn as having the very best capability in the field of radiochemistry, that the separating out of radium was impossible for him. It can be said in Hahn's defence that the production of pure mesothorium is not realisable because *radium and mesothorium have the same chemical properties.*

The same chemical properties, and yet they are not identical? For regarding their radioactive constants, radium and mesothorium can be identified as different. Who could solve this puzzle?

Just as Professor Boltwood had had to acknowledge Hahn's performance in the discovery of mesothorium, the mother substance of radio-thorium, and more fairly congratulate him for his success, things were reversed in the search for the mother substance of radium. Also, both researchers were setting out from the same compelling idea, that the time it took for radium to decay to one half in quantity was around two thousand years. Hahn had already concluded it was so in 1907. *In the far distant past there was radium in the solid crust of the Earth; also if the entire Earth were to have been made of it, today it would have long since vanished if we were not able to make another assumption, namely, that radium itself is constantly formed anew.*

But out of which element would radium be formed? Since it always occurs coexisting with uranium, it would have to have been formed from this

---

[1] *'Über die Eigenschaften des technischen hergestellten Mesothoriums und seine Dosierung'.*

element. This was first suspected by Hahn and Boltwood. Despite many an attempt they could not detect any newly formed radium in radium-free samples of uranium even after prolonged periods of observation.

Boltwood seized upon a new hypothesis, that radium descended from ac-tinium. However, Hahn did not believe that was so, but wrote to Joachim-sthal in Bohemia and requested a particular fraction from the processing of uranium pitchblende in which he believed he would have to seek the progenitor of radium, which he did, indeed, also find. But this time he lost the scientific contest with his old rival.

After a year long search Hahn was just a few weeks too late. His paper 'On the Mother Substance of Radium', the manuscript of which he com-pleted on 17 October, the *Berichte der Deutschen Chemischen Gesellschaft* published in November 1907. A few days after Hahn had put his paper in the post he fell upon the October issue of the *American Journal of Sci-ence* with Boltwood's 'Notes about a New Radio-Active Element'. In it the American described his earlier result and described the successful finding of the element sought, and which he called 'ionium'. Boltwood had taken practically the same route as Hahn, only he had been quicker.

5.2   It's Unbelievable What Qualifies as a University Lecturer Nowadays

Recognition engendered willingness in the father to open doors for the son's university career by financial gifts. Apart from a fixed monthly allowance Heinrich Hahn also paid various accounts and showed understanding for petitions like the following. *The purpose of this letter is to ask you to send me one hundred and fifty marks. I have bought myself two milligrammes of pure radium for one hundred and fifty marks. The list price is six hundred marks, and I thought that advantage ought to be taken of the opportunity . . .* (15 May 1907).

That such investments were worth it, Otto Hahn was proudly able to report to his father a few weeks later on 3 July, *. . . from a local scientific foundation I have received two thousand marks for the furthering of my work on radioactivity. . . , in this respect it is very good to me, as it means and shows a certain encouragement that the many radioactive studies are not viewed as sheer fantasy.*

In January 1907 Otto Hahn applied to the Faculty of Philosophy of Berlin University for the authority to teach as a Privatdozent[2] in chemistry. The Faculty accepted Hahn's paper on radio-thorium in the *Jahrbuch der Ra-dioaktivität und Elektronik* of 1905 as his habilitationsschrift[3], which he had

---

[2] *Translator's Note:* An unsalaried lecturer.

[3] *Translator's Note:* The qualifying publication for a first post as a lecturer.

submitted together with ten other publications. As expert referees he had at his disposal Emil Fischer and Walter Nernst. Professor Nernst became known through his theoretical works in the field of physical chemistry. He had made a name for himself through his invention of the so called Nernst light bulb.

At that time applications for approval of the habilitation procedure were publicly displayed on notice boards in the Faculty. As well an announcement was circulated which all representatives of the teaching bodies had to sign.

It speaks for the prejudice against a substantially unknown area of research, but one rich in new insights, that Hahn's habilitation application was annotated with many disparaging remarks. "It is unbelievable what qualifies as a university lecturer nowadays", noted one of the colleagues.

Emil Fischer's verdict on the achievements of his candidates was, on the contrary, always positive. He gave expression to his persuasion that "Dr. Hahn is to be completely trusted with refined methods of radioactive research and has the ability to use them for the obtaining of new, finer results. Therefore it seems desirable to me", wrote Fischer further in his report, "that this promising branch of physical research should be cared for here more than it has been, therefore I would like to establish Dr. Hahn in the Chemistry Institute, and for the same reason I hold his habilitation to be desirable. Since he has fulfilled all the pre-conditions I therefore hold that we can not have the least reservation against his application for registration for a trial course of lectures".

For this trial lecture course Fischer selected from Hahn's proposals first the theme 'The Modern Concept of the Constitution of Matter'. But then he decided on a harmless theme, 'The Most Important Salts of the Earth'. As his public inaugural lecture Otto Hahn spoke about 'Current Problems in Radio-Active Research'.

In the university files, which fortunately still exist, and to which we owe this information, there is also an assessment delivered by Nernst. Professor Nernst did not appear to be quite so convinced of the capability of Otto Hahn: "... that the candidate thoroughly acquainted himself with the methods of radioactivity research in the laboratories of Ramsay and Rutherford and understood how to implement them with all expertise, the works presented leave no doubt. I must admit, however, that I am not quite so sure what the capability of Dr. Hahn is in conducting original research independently; there is nothing to show that up to now he has not merely worked under the direct guidance of the abovenamed researchers, therefore I would have liked to have seen whether the candidate could also demonstrate that something in the work carried out was done on his own initiative... ".

To seek to be habilitated at the University of Berlin was not finally a question of money, as may be gathered from the letter of Otto Hahn of June 15 to his father. "*I send you two receipts herewith, for you to see how one disposes of one's money here. The one hundred and fifty one marks I paid*

when I applied for the registration for habilitation .... The eighteen marks
are a small example of how one must shell out in the process of habilitation
here. In addition: 1. Caretaker—5 marks for drink; 2. Caretaker—5 marks
for drink; etc.., etc.."

The examination procedure took place on the same day, on 15 June
1907. In connection with his habilitation hearing Hahn had a critical situation to weather. Inevitably the discussion turned to radioactivity. Emil
Fischer thought, shaking his head, that he could only imagine with difficulty the imponderability of the methods of measurement in radioactivity,
detecting quantities of substance usually of the order of magnitude of 10^{-12}
grammes or less. He, Fischer, did not want to believe this was right. The
most sensitive test for certain organic substances was always the sense of
smell, therefore one's nose .... With this sensory organ one could register
certainly minimal, even almost vanishing concentrations.

Fischer had in that connection once undertaken an unusual experiment.
In a large hall he left hidden some methylmercaptan, the dreadful smell of
which was always perceptible in the slightest quantity. Starting with small
quantities, he increased the concentration. At his instruction investigators
entering sniffed about the room and confirmed or rejected that there was
something to smell. Fischer, deadened to the infernal stink, could no longer
be counted as an impartial witness. From the cubic capacity of the room
and the quantity of mercaptan sprayed, Fischer calculated a limiting concentration which actually laid down at that time the limits of detection of
the familiar analytical methods.

5.3   An Element Is Not an Element

Emil Fischer's slight criticism was not taken in earnest, for he had also
made demands on and supported Otto Hahn's work. On the other hand
Hahn often came to feel that his work, and especially the significance of his
researches in radioactivity, were held in doubt, even that it was sought to
bring them into discredit. Something drastic had to be done, so he would
make vividly clear the 'dilemma' to a few scientists of the older generation,
with a detailed description. The place of the symbolic act was to be the
main gathering of the German Bunsengesellschaft for Applied and Physical
Chemistry in Hamburg. The agenda of this gathering in May 1907 is a
unique contemporary document.

10 May fell on a Friday. In the large lecture theatre of the State Laboratories the rows of seats quickly filled that morning. The Privy Councillor[4]
Professor Nernst from Berlin opened the main meeting. The theme of the
day—Radio-Activity and the Atomic Decay Hypothesis. Nernst indicated

---

[4]See Note 8 in Translator's Notes on the Text.

that the lecture ought to end at one o'clock so that there could be a general debate about the matter. This round of discussion everybody viewed with great interest.

A convincing example for the effect of radioactive rays was provided by the Viennese radium researcher Friedrich von Lerch in his lecture. He reported a method of precipitation of the radioelement radium C and said, "The quantity of the substance RaC which is needed to discharge an electroscope in one second is found to be 10^{-10} grammes... If we distributed one milligramme of RaC amongst all people living on the earth—some two billion—the quantity of the substance that each person had in hand would make five electroscopes' leaves collapse together in a fraction of a second".

Nernst appeared to be impressed.

Afterwards a paper was given on the atomic decay hypothesis. Admittedly the name of the lecturer, Dr. Hahn from Berlin, was known only to a few, but everyone had heard something about the theory of radioactive decay, and there was inconsistency in what was circulating. Nernst referred to the announcement that Hahn had worked with Rutherford, and that one could expect that his information would be first hand.

Dr. Hahn explained briefly the phenomenon of radioactivity and acknowledged Rutherford's achievements. In order to illustrate the problems of applied radioactive research, Hahn described Boltwood's investigations with the help of the radioactive uranium–radium balance in estimating the age of minerals and the Earth. Boltwood had found the same ratio in various uranium minerals, and indeed the inconceivably small concentration of 0.000 000 38 grammes of radium per gramme of uranium.

With a few numbers the lecturer began to awaken his listeners to the immense energies that lay within the atom. Hahn commented on the example of the *energy which a gramme of radium gives off during one year. It is many thousands of times greater than that which would be obtained by the explosion of the same amount of the most powerful explosive.*

What could lie closer at hand than making this inexhaustible source of energy exploitable? Otto Hahn—*At such levels of energy our chemical and physical supplies can not approach such levels of energy, at least not for the time being.*

Privy Councillor Gustav Tammann, Professor of Inorganic Chemistry at Göttingen, an internationally recognised scholar, opened the afternoon with a question the discussion session awaited with suspense. He belonged to the generation of those scientists who stood helpless before the latest discoveries in the field of radioactivity and atomic structure. So it was no surprise that he met Hahn's explanations with mistrust. Tammann as always, and for which his pupils praised him, displayed a "natural judgement of rare sureness of aim" .... Of Hahn's report about the detailed uranium–radium ratio he remarked, "I think the theory demands more. You not only demand that the ratio is constant, but also that the distribution of the radium within the uranium is perfectly regular and homogeneous. That is, as far as

I know—and there are people here who follow the literature much better—is not detected with complete accuracy. If, however, it is so, it means that the estimated analysis performed by Dr. Hahn is hardly secure".

With the reasonable objection that a constant uranium–radium ratio was only a preliminary assumption, that is to say, minerals found in nature would be expected to be unchanged since ancient times, Otto Hahn sought to convince the discussion's speaker of the opposite. But without success, as will be revealed.

Tammann continued, "In Boltwood's work has the constancy of the ratio of radium to uranium been shown?"

Hahn replied, "*Certainly. Boltwood has tested quantities of substance between one half and one gramme, ... as far as I can judge, about forty different uranium minerals. The result in all cases was absolutely the same*".

Tammann rejoined, "Therefore the question appears to be open. What has been proved for certain is nothing up to now. Boltwood had had to have found by precise tests in various uranium minerals an exceptionally different distribution of radium".

"That is a misunderstanding!", volunteered Professor Ferdinand Henrich from Erlangen, indicating his wish to speak. He was one of the experts and had himself given a lecture on this theme.

Spiritedly and quite disrespectfully, Otto Hahn contradicted, for after Rutherford's colloquia he could not help it. A few older scientists were visibly annoyed by that, for Hahn's manner was not fitted to the traditions which the academic hierarchy expected of him. "What kind of person is that?", would be asked, and someone would answer, "An anglicised Berliner".

Once again Tammann took up speaking. "Another question that I ought to bring into discussion, for nowhere else will one be so quickly and surely enlightened as here ... ". The ironic undertone was not noticed. "I get the impression that it has been said many times today that the emanation belongs to the inert gases. I can not quite agree with that, for up to today nothing has been proved about the well known inert gases to show that they somehow decay and are links for us, and that they are not elements. This touches on the question, are the radioactive elements actually elements, gentlemen? Radium fits into this, as we know, yet is not in the periodic table... But we shall probably be instructed better in a short time".

In the general tumult that then broke out in the lecture theatre Tammann's last words were hardly understood. There was a bedlam of discussion between everyone. Incensed interruptions were loud, but also applause and much laughter was heard to break out.

In his capacity as chairman Nernst finally obtained some quiet after many appeals, and sought to settle the quarrel with the wisdom of Solomon.

"It is all a question of definition. One can certainly make the definition that an element which stays constant in its mass is an element, and that an element which alters itself radioactively is not an element".

Nowadays we know that such a starting point is untenable. The scientists present at the Bunsen Meeting also could not warm to this Nernstian definition. Objectivity only returned again to the discussion when Otto Hahn asked to speak once again. *I must first make a reply to the question about the nature of radioactive emanations. ... In general one calls inert gases those which up to now have not been brought into any combinations by even very energetic reactions. The emanation of radium was conducted over red hot magnesium, red hot copper, through the various reagents which had acted on all the other gases which did not belong to the group of inert gases, in order to cause a reaction. The radium emanation was found to have been unchanged after it had passed through all these systems... .*

"I still would not count them amongst the inert gases", Tammann interrupted the speaker, "because the inert gases do not undergo radioactive decay".

The question of the differences between the radioactive emanations and the rare gases, went on Hahn immediately without any hesitation, *stands and falls on the second question of Professor Tammann, of whether radium is an element. ... Radium has been up to now, and always will be by the great majority of researchers, regarded as an element, even though it emits rays. The difference between itself and other elements is only slight. Uranium has always been regarded as an element and also emits rays. We have elements which decompose in three seconds, and we have some that decompose in a thousand million years, like thorium and uranium.*

Professor Bohuslav Brauner from Prague offered for this point a strange theory. "I imagine things like this, that it is in fact already dead, and extinct elements can yield elements, which no longer exist".

A little mockingly, Nernst said, "The hardly comforting hypothesis of colleague Brauner that there are already extinct elements also offers one the opportunity of counter-posing the joyously hopeful hypothesis that some individual elements have not yet been born".

In this humorous turn there lay a tiny seed of truth in the future. At that time there were places in the periodic table for only ninety two elements. A few of them had not yet been discovered. That gave rise to some gaps. Today we know one hundred and nine elements, of which those with position from 93 to 109 are actually made artificially. Also the basic chemical substances at positions 43, 61, 85, and 87 were not known in 1907 to occur in nature, but would first have to be 'born' in the laboratory.

In the lively discussion in the Hamburg Bunsen Meeting, a serious scientific problem was brought into the arena. For all the enthusiasm which the discovery of a new basic chemical substance always revealed, the finding of so many radioactive elements had finally given rise only to perplexity and bewilderment. The reason was that the radioelements no longer could be categorised within the periodic table. Only the first discovered elements, uranium, radium, polonium, thorium, and actinium had found their proper locations.

"I am very worried about what ought to be begun to be done in the periodic table with the many radioactive elements... ", yielded Professor Brauner during the Bunsen Meeting. Otto Hahn himself, who put his trust in the latest theories of Rutherford and Soddy, had to concede that it was impossible to order all twenty five radioelements found hitherto according to the periodic table.

Could the system of Mendeleev and Meyer, conceived with the vision of genius and many times since tried and tested, have lost its validity in the case of radioelements? Or were those new radioactive substances actually not elements, perhaps?

5.4 Lise Meitner

Towards his colleagues at the University of Berlin, many of whom held ambition in their careers, Otto Hahn showed no rivalry after his habilitation. *Radium will not be taken seriously by the chemists*, regretted Hahn. For that reason the young privatdozent[5] in chemistry sought contact with the physicists at the university. In his opinion physics offered more for his special area of work than chemistry. On these grounds—as Hahn justified to himself—physics had always been his love, even if an unrequited one.

Thus do we soon find the radiochemist Hahn a regular visitor at the weekly colloquia of the Physics Institute. Amongst the young physicists he soon counted Otto von Baeyer, James Franck, Gustav Hertz, Peter Prings-heim, and Wilhelm Westphal as his friends. All became later on well known, even famous, scientists. Two of them—Hertz and Franck—won the Nobel Prize for Physics. In this circle of friends, on 28 September 1907, Otto Hahn made the acquaintance of the physicist Lise Meitner, who was the same age as him. That day was to become a historic date in the history of research in the natural sciences.

Lise Meitner was Austrian. She hailed from Vienna, where she was born on 7 November 1878 the daughter of a Jewish lawyer. Early on the young girl showed herself to be above average, matriculating in 1901 with distinction, and then studying physics for eight semesters at the University of Vienna and taking her doctoral examination in February 1906 on the topic 'On the Conduction of Heat in Inhomogeneous Bodies'. She was the second woman to have obtained her doctorate as a physicist at the University of Vienna.

Crucial to Lise Meitner's scientific career had certainly been her collaboration with Stefan Meyer, who later on was the Director of the Vienna Radium Institute. Under his direction she took her first steps in the field of atomic physics. Two publications from this time bear testimony to that,

---

[5] A lecturer depending upon a private income.

'On the Absorption of Alpha and Beta Rays'[6] and 'On the Scattering of Alpha Rays'[7] appeared in the *Physikalische Zeitschrift* of 1906 and 1907.

Amongst the mentors of the young physicist were also Ludwig Boltzmann and Franz Exner. Boltzmann's enthusiasm and his lively presentation of theoretical physics left a very long lasting impression upon Lise Meitner. She was deeply affected by Boltzmann's sudden decease in 1906. His death made up her mind to leave Vienna in order to work under Max Planck in Berlin to complete her training.

"As far as Planck's lectures were concerned", Lise Meitner recalled in 1958, "I must confess that at the start I was almost disappointed. I was a pupil of Boltzmann, and Boltzmann was enthusiastic about his science, which quite naturally carried us young students away". From this point of view the Vienna lady physicist must have found the clear presentation of Planck's lectures impersonal, almost a sobering experience. However, she soon recognised that behind the solidly factual expositor there stood a committed and imaginative scientist. Both began to have a regard for each other, and Lise Meitner even became a guest the Planck family was fond of seeing.

Lectures and seminars did not fill the daily routine of the young researcher. She wanted to do experimental work again. During a colloquium in the Physics Institute, which in those days took place in the form of a small circle within the library, she asked the Institute's Director, Professor Heinrich Rubens, if he could help her in that respect.

Privy Councillor Rubens looked around the circle of the gathered physicists and well saw that the almost timidly made request of Dr. Miss Meitner would be received sceptically. "I only really know two with whom you can work. Dr. Ladenburg, a physicist in my Institute, and Dr. Hahn, a chemist and specialist in new methods of radioactive research." With a mischievous smile, Rubens asked those present, "Who of the two gentlemen should have Miss Meitner?"

When it was seen that the young lady was embarrassed by this question Otto Hahn helped her out. "Let us allow our colleague to decide for herself which field of work appeals to her". Lise Meitner chose to work with Otto Hahn. To step out into a new scientific land appealed to her very much. *So she came to me*, wrote Otto Hahn in his autobiography, *and out of two years spent in Berlin came more than thirty years of collaborative work and lasting friendship*.

Although the start of their practical work was bound up with difficulties, because the Prussian university statutes in practice allowed women no possibility of an academic career, the Director, Emil Fischer, who in this matter was very conservative, permitted Lise Meitner to have access

---

[6] *'Über die Absorption der Alpha- und Betastrahlen'*.

[7] *'Über die Zerstreuung der Alphastrahlen'*.

only to the wooden workshop and an adjoining space. She was not allowed to enter the students' lecture theatres and laboratories. But at the end of 1908 a law submitted to the Prussian State became effective, which allowed women also to study at university. From then on Hahn's lady colleague was able to move freely about the institute.

In October 1907 the pair began their research collaboration. They were able to work as a team in an exemplary way, the performance of which caused a stir in a short time. A wealth of unsolved questions in the area of radioactivity awaited the pair. Which should they solve first? Both held it to be right first to study the beta rays of the newly discovered meso-thorium. Important physical details such as penetration and speed should be ascertained. On 8 August 1908 they completed the manuscript of their first joint publication 'On the Absorption of the Beta Rays of Some Radio Elements' [8]. The *Physikalische Zeitschrift* published the article in its issue of 15 May. This publication was the prelude to fifty joint works, all in all.

As it later turned out, the conclusion to which the two radium researchers had come in their first work was not valid. But these investigations were not in vain. On the contrary, they led directly to the finding of new beta ray emitting radioelements, which since then have been known to experts by the mysterious sounding names of Ac C'' (actinium C''), Th C'' (thorium D), and Ra C'' (radium C_2). These new substances belong to the actinium, thorium, and radium decay sequences, respectively, and up to then had either not been discovered or had simply been overlooked.

Through these investigations Otto Hahn found a new effect which he named 'radioactive reaction' and which Rutherford had already predicted in 1904. Together with his colleague, Otto Hahn developed an elegant method of preparation with the help of which one learnt to separate out and identify imponderable radioactive products, and which is still used today.

The finding of the reaction effect is again an example of Otto Hahn's subtle method of working. As usual, a result can only be viewed as certain if all measurements are meaningful and give reproducible values, and un-deterred he sought the cause of inexplicable deviations, which others liked to dismiss as 'the effects of contamination'. By the separation from the decay products of actinium this radiochemist brought to light a strange activity one hundred thousand times weaker than that of the isolated test substances. Since his posts with Ramsay and Rutherford, Hahn preferred to go after such anomalies. Perhaps an unknown element might lie hidden there. Otto Hahn lifted the veil from this secretiveness. The remaining ac-tivity arose not from a new substance, but from the rest of the actinium atoms, because these *suffered a reaction through alpha ray bombardment, similar to that of cannons, if the projectile departed from its course.*

---

[8] *'Über die Absorption der Betastrahlen einiger Radioelemente'.*

Looking back on it, Otto Hahn thought that he had been rash to have somewhat hurriedly—on 16 January—put the manuscript in the post and to have given the paper the somewhat misleading title 'On a New Phenomenon in the Activation of Actinium'[9] (published in the *Physikalischen Zeitschrift* 1 February 1909).

As the selfsame Lise Meitner's eyes lit upon the manuscript of her colleague she concluded, with her own special keenness of intellect, that the effect described was suitable for building upon. Hardly a week later, on 22 January 1909, Meitner and Hahn reported at the Physikalische Gesellschaft in Berlin further possible applications of the reaction effect discovered by Hahn, giving a talk entitled 'A New Method for Making Decay Products'.

During the colloquium in the Physics Institute, the pair both reported their newest research results. Then concepts buzzed through the air, such as thorium X and thorium D, radium C, radio-actinium, mesothorium 1 and 2, as well as the numerical values of radioactive and chemical constants, in a bewildering sequence throughout the room. The Director, Rubens, himself shook his head and then gave up: "How is it possible for you to distinguish all these names and also to know their chemical properties? It is all fearfully complicated".

Hahn had to smile at such remarks. A physicist could lecture in no other way. A matter of satisfaction to him was an admission of his friend the famous physicist Max von Laue, who received the Nobel Prize for Physics in 1914 for his discovery of Röntgen radiation interference. Around fifty years after these events von Laue recalled a popular lecture of Hahn: "It was the first time, dear Otto, that I really understood what you were actually doing. Your business was never understood properly... ". For all the trouble which Meitner and Hahn spent on the development of their work, they could not make an efficient place for research out of the wooden workshop. For this reason they sought the advice of Otto von Baeyer of the Physics Institute. Thanks to his support, many investigations could now be speedily pushed forward, above all the very necessary determination of the beta ray spectra. Joint publications testify to their fruitful collaboration. The wide path between the wooden workshop and the Physics Institute frequently conveyed problems. Some of the separated radioactive products were so short lived that Hahn and Meitner had to be quick about their business. Usually they fixed these radioelements in invisible 'amounts' on a simple wire. Then the two ran, Meitner in front and Hahn behind her, with their test to the exit of the Chemistry Institute, to the general amusement of colleagues and the pedestrians passing by on Hessische Strasse. There waited *a car*, told Hahn with amusement, *which then took us to the Physics Institute on Reichstagufer where Baeyer had prepared everything. It did not always work....*

---

[9] *'Über eine neue Erscheinung bei der Aktivierung von Actinium'.*

Lise Meitner was always pleased if Otto von Baeyer, James Franck, Gustav Hertz, Wilhelm Westphal, and the other physicists turned up at the wooden workshop for a return visit. They seldom chose the official route for arriving at Hahn and Meitner on the ground floor. It was a carefree time which Lise Meitner liked to recall: "We were young, happy, and carefree—perhaps politically too carefree".

5.5   Isotopy—The Puzzle's Solution

Are our radioactive substances actually elements in the chemical sense? This ticklish question about each newly found radioelement was posed by Lise Meitner and Otto Hahn. There was no other question for many departmental colleagues. No way had been found up to then of incorporating the radioactive elements into the periodic table. Even such outstanding scientists as Rutherford, Soddy, Niels Bohr, and Marie Curie had not found an explanation in the interim. An additional complication was that most of the newly found radioelements disintegrated quickly and could always only be isolated in immeasurable quantities. The determination of their atomic weights or even their chemical properties, a prerequisite for their classification, could not even be contemplated for that very reason.

As one of the speakers at the Bunsen Meeting of 1907, those days saw Otto Hahn considerably confronted by this tiresome problem. In the meantime the situation had become more conspicuous than ever. When Lise Meitner reported on the decay products identified in her joint work with Hahn at the meeting of German natural scientists and medical practitioners in Salzburg in September 1909 about the latest news, a similarly heated debate threatened, as in the Bunsen Meeting two years before. In view of the number of radioelements having grown to a considerable number, during the discussion Privy Councillor Rubens thought, "It is certainly very nice that the radium family has had more happy additions. But as time goes by that will become somewhat eerie, and one will ask oneself whether the increase will be able to continue further... ".

Only a new theoretical foundation could bring clarity. The solution was delivered a short time thereafter—its first beginnings were already in that year of 1911, and its final formulation was given in 1913 by Frederick Soddy with his theory of isotopes of elements. According to it, a single element exists in several atomic forms, namely isotopes, which have various atomic weights, also called the atomic number. Many elements are pure elements existing in only one atomic form with a well defined atomic weight. Mixed elements, in contrast, exist together with various heavy isotopes. Isotopes of an element are chemically indistinguishable from each other, and therefore cannot be separated from them. They possess, however, specific physical differences, which in radioactive elements are exhibited in the kinds of decay and characteristic half life. Of course, the determination of the atomic weight would now no longer suffice for allocating an element its place in the periodic table. Firstly, with the introduction of a further identification quantity for each element, the ordering number or atomic number actually introduced an 'ordering'. However, it remained unclear why isotopes of a single element could have various masses. These questions first found an explanation twenty years later.

The new theory, which was soon confirmed experimentally and augmented by the so called radioactive shift periods, at one blow explained the well known problems. All the radioactive elements discovered in the early days turned out to be varieties of chemical elements already well known. Only very few representatives were actually to emerge as entirely new elements, which consequently ought also to be entitled to a place in the periodic table. Radioactive emanations were none other than isotopes of the inert gas radon. Hahn's radio-thorium was a thorium isotope with an atomic weight of 218, and mesothorium, likewise discovered by him, a radium isotope of atomic weight 228. Both were therefore not new elements— a forgivable mistake, which should be blamed on the incomplete atomic theory of those days.

Now Otto Hahn could explain the many vain attempts he had undertaken to separate, for example, radium from mesothorium or thorium from radiothorium. These projects simply could not succeed because each was a matter of one and the same element. But to his undoubted regret, Otto Hahn had yet again established later that on the basis of his laboriously painstaking and detailed work he had acquired knowledge earlier than Soddy about the same thoughts, but had shrunk from clothing them in a theory or publishing them. *Soddy,* so Hahn thought, *surely has not made as many negative attempts at separation as I, but he has more courage.*

5.6 International Meetings

The research examples found in their joint work brought Lise Meitner and Otto Hahn to international recognition. The Berlin researchers were counted as welcome guests at scientific congresses. The pair had close contact with noted radium researchers of the entire world which many times turned into lifelong friendship. Since at that time only a few researchers in the individual Lände had made their mark in the new research area, one quickly found common ground, and one was soon confided in about the work of another. With good reason, an 'international family of atomic researchers' was spoken of at that time.

The 'founder' of this family, the doyen Rutherford, received the Nobel Prize in December 1908. This award brought pure joy to his pupils and friends around the world. When Rutherford travelled to Stockholm to receive the Prize he was besieged with invitations. Otto Hahn's plea for him to visit Berlin on his journey back Rutherford could not refuse. It turned into a warm reunion.

"Do you know that I am doing chemistry extraordinary?", Rutherford said immediately and insistently in his booming voice to his former pupil, scarcely allowing the latter to congratulate him on his Nobel Prize. Hahn attempted a diplomatic turn, "Your strengths really lie in the physics arena... ".

"Nonsense! They've given me the Prize for Chemistry—do you hear that? Chemistry! To me!! A full blooded physicist who has never done a single experiment in chemistry! And the devil take me if I ever so much as touch it".

In fact, this researcher had been cited "for his researches into the decomposition of elements and the chemistry of radioactive substances"—a curious circumstance, for Ernest Rutherford had investigated and interpreted radioactivity purely physically.

For his host Rutherford had a surprise ready—an invitation to the meeting of the British Association for the Advancement of Science taking place in Winnipeg in the following year. Hahn accepted the invitation with thanks. The journey in that hospitable country, which he took with a colleague in August 1909, remained one of his fondest memories. During the meeting Rutherford gave an enthusiastically attended lecture on 'The Latest Progress in Atomistics'. Hahn reported on 'On the Production of New Radio-Active Decay Products'.

In September 1909 he went back to Europe with Rutherford. To the compelling pleas of his distinguished mentor, Otto Hahn interrupted his journey home in order to accompany Rutherford to Manchester, where that physicist had worked for the last two years at the university.

Hahn and Rutherford were soon to meet again; to be precise, at an international 'Congress for Radiology and Electricity' which took place in Brussels from 13–15 September 1910. At the same time there was a reunion with colleagues from around the world. During the congress Hahn made first acquaintance with Marie Curie and Frederick Soddy.

The congress in Brussels had been convened in order to clarify a few current problems in the stormily developing field of radium research. Questions of a standardised nomenclature and internationally agreed standard were to be considered. Marie Curie and Ernest Rutherford, the chairman and chairwoman of the congress, suggested the choice of an international commission which should be made up of the most prominent radium researchers: Marie Curie and Debierne (France); Rutherford and Soddy (England); Boltwood (USA); Arthur Eve (Canada); Stefan Meyer and Egon von Schweidler (Austria); as well as Hans Friedrich Geitel and Hahn (Germany).

The French researcher said that she was ready to produce a radium standard of high purity of about twenty milligrammes. The congress passed with unanimous approval a resolution that the unit of radioactive radiation should henceforward be called the 'curie'. In her modesty Madame Curie wanted to decline. Finally she accepted the suggestion with the hint that she regarded it as an honour for her husband, Pierre Curie, who had been the victim of an accident. One gramme of radium would henceforward be the measure of one curie of radioactivity.

When the International Radium Standard Commission assembled once again—this time in Paris, from 25–28 March 1912—Madame Curie presented 21.99 milligrammes of high purity radium chloride. Melted down inside a small glass tube, protected by thick lead walls, secured like a superb treasure in a safe, since then this radium standard has been kept in Sèvres near Paris at the international 'Bureau for Mass and Weight'. However, this was not because of its considerable market value. This radium standard is much more to be compared in its absolute significance with the unit of length, the standard metre, which is also kept at Sèvres. For individual states reserve standards were made ready, which would be deposited in the participating countries, which in Germany was in the Physikalisch–Technische Reichsanstalt in Berlin.

To Lise Meitner Otto Hahn complained that his work for the Radium Standard Commission left him so little time over for sightseeing the Seine metropolis. His coworker showed hardly any understanding for that, as she wrote to Hahn on 9 April 1912, "I am already curious to know what you have to say about Paris. Finding yourself having to do all sorts of things, does it not annoy you that in vain is one not famous?"

5.7   The Nucleus of the Atom

During his stay in Paris in March 1912, Otto Hahn strengthened his acquaintance with Marie Curie, the discoverer of radium. As a guest in her home he got to know both her daughters, Irène and Eve. Hahn vividly remembered their masterly piano playing. Marie Curie loved to listen to the compositions of her Polish countryman Chopin.

Rutherford had let nothing stop him from inviting the members of the Radium Commission to a meal in the Café de Paris, which was numbered amongst the most exclusive of the restaurants of that city of the Seine. Despite the outstanding menus, Rutherford seemed not to be satisfied. The old grouch was not comfortable, having to sit squeezed in so tightly at the table. After the meal he proclaimed himself relieved and beckoned Otto Hahn over to him in order, rising and departing, to begin a conversation with him. The reproachful looks of the distinguished guests in the restaurant did not bother Rutherford.

"I now know what it looks like… ", began Rutherford.

"What?"

"… the atom!"

Contrary to his custom, the physicist went on further to make it plausible to Hahn what an atom was. For a few years it had been compared to a massive billiard ball. After the discovery of the electron it had been thought of as an electrically neutral object, on the surface of which released electrons had settled.

Otto Hahn thought he knew what his mentor was aiming at with his digression. In the previous year Rutherford had written an article for the *Philosophical Magazine* in which he attributed to the atom a 'central charge'. Hahn received an offprint and thought that this theory certainly was not especially original. It seemed of no significance to him whether the electric charge was distributed over a spacially diffuse structure or centralised somewhere.

"No, dear Hahn, that is not it. I now know what the atom actually looks like. The atom has a nucleus".

An atomic nucleus? Yes, that was something new! With eloquent words the atomic physicist explained his experiments which had driven his daring supposition. A platinum foil had been bombarded with a shower of alpha particles. Rutherford could show that one particle in eight thousand which pelted the foil was deflected, and even had been thrown backwards. But what was it that could stop a particle with high characteristic mass which was moving with a speed of fifteen thousand kilometres per second through an atom's space. It could only be an obstacle that was more compact than the alpha particle, and tiny as well, which could be hit only extremely seldom—namely the nucleus of the atom.

Hahn had to admit that Rutherford's train of thought seemed tempting. The existence of the atomic nucleus made possible the explanation of many previously almost inexplicable phenomena. The nucleus, which had to be the location of a prevailing charge pressed together in a ball in which the total mass of the atom would have to be concentrated.

A more precisely stated theory about the existence of the atomic nucleus was published by Rutherford in the *Philosophical Magazine* in August 1912.

The Danish physicist Niels Bohr, who had become the leading theorist in the area of atomic theory, took hold of Rutherford's concepts and commented on them in 1913 in more papers with his ideas with the help of a new type of atomic model. In that, the atom consisted of a positively charged nucleus which combined the entire mass, shrouded by negative electrons which described certain orbits.

This form for the atom was now accepted. Admittedly, it took a little time before concrete information about the constitution of the atomic nucleus could be obtained, but valuable theoretical discoveries were already being derived, such as the isotopy of chemical elements and their insertion into the periodic table in accordance with the charge of their nuclei, through which the laws of radioactive decay became understandable. The nucleus alone came into question as the source of radioactive rays, and as the seat of the enormous energy of the atom. In contrast, the electron cloud residing around the nucleus was responsible for the absorption and emission of light and Röntgen rays, and for the ability of an atom to have chemical reactions.

Now scientists had a clear idea of the scale of things. *En masse* people measured the diameter of an atom and arrived at a value of 10^{-8} centimetres, one one hundred millionth of a centimetre. The measurement of the nucleus still seemed inconceivable, which was estimated to be ten thousand times smaller than the whole atom. Who, however, outside the specialists, could describe it pictorially? Here the every day projection model would have to help in making these dimensions of the microscopic world comprehensible. Hahn chose for this purpose a vivid example. In its size the nucleus compared to the whole atom rather like a pinhead does to an average sized house. Actually one would have to say that the atom is empty, for the entirety of the matter resides in the nucleus, even that very pinhead. Now one had only to learn the trick of making estimates correctly, which served experimentalists like Rutherford, in order to seek out the famous pinhead, the atomic nucleus, in the haystack.

5.8   An Absent Minded Professor

The choice of Otto Hahn in scientific committees, such as the International Radium Standard Commission and later in the International Atomic Weights Commission, was doubtless an expression of the high regard for his professional qualities and personality. It was not accidental that at this time he was appointed professor. In the 'Professorial Patent', as it is called, prepared by August von Trott zu Solz, the Prussian Minister of Ecclesiastical, Educational, and Medical Affairs, on 10 October 1910, in the characteristic cursive script of the Kaiser's Germany, "Therefore in consideration of his acknowledged scientific achievements, to the Privatdozent Dr. Otto Hahn of the Philosophical Faculty of the Friedrich Wilhelm University of Berlin I have awarded the title of 'Professor', and grant the same the present patent on the premiss that the hereinstated Professor Dr. Otto Hahn remain devoted with inviolable loyalty to His Majesty the King and the highest of houses and to the forwarding of science as he has heretofore concerned himself, and in public recognition and protection by the title I am pleased to lend him, ... ".

Evidently, with this appointment Otto Hahn made his own the proverbial absent mindedness of a professor. In May 1911 he went in the Baltic direction in order to take part in a meeting of the Society of German Chemists in Stettin. He wanted to give an experimental presentation on the theme 'Properties of Mesothorium and Radiothorium'. For this purpose there had been made under his direction in the Knöfler thorium plant a mesothorium preparation of a few hundred milligrammes. The precious demonstration item had a value of over 100,000 marks. Mesothorium in this amount gave off strong radioactive radiation, which did not prevent the Berlin radium researcher from lugging it around with him in a simple briefcase.

Professor Hahn travelled to the meeting place by tram. With his thoughts obviously already on the lecture, he forgot the briefcase when he got off. Only by a daring leap onto the tram already in motion could he rescue his preparation and thus his lecture.

At the conclusion of the meeting the organiser invited everybody on a steamer trip. Here on the waves of the Bay of Stettin, and not in his work, Otto Hahn was to make the 'discovery of his life'. On board he got to know a young lady, who first caught his eye despairing for her fashionable summer hat as she sought to rescue it in the fresh sea breeze.

The twenty five year old Edith Junghans, a student at the Royal School of Art in Berlin, was spending time in Stettin at the wish of her parents. Her father, Justice Councillor Paul Ferdinand Junghans, as the appointed city chairman, was one of the worthies by whom the distinguished guests had been invited. Hahn's absent mindedness also played a trick on him with his new acquaintance. He forgot to enquire about the address of the young lady.

After the return journey from Stettin to Berlin the young scientist could not get his acquaintance out of his mind. He had to confess that this discovery of a moment engaged him more strongly than all the radioelements together. Otto Hahn, now thirty three years old, seemed firmly resolved not to let this connection come to an end. Back in Berlin he quickly resolved to write a card to 'Fräulein Edith Junghans, The Royal School of Art, Berlin'. The post reached its target and contact was re-established to mutual delight.

And Lise Meitner? His relationship to her, his closest coworker, was always warm. In his memories Hahn stresses that again and again. Of Lise Meitner, whom he characterised as *very restrained, almost timid*, he wrote in recollections of his life, *Of points in common between us outside the Institute there is nothing to say. We never even went out for a walk together. Apart from colloquia we met only in the wooden workshop. There we usually worked until shortly before eight o'clock (in the evening). Lise Meitner went home and I went home. At the same time we were warm friends.* Lise Meitner remained unmarried.

On an autumn day at the beginning of October 1912 Otto Hahn and Edith Junghans took a walk to Berlin–Dahlem. Hahn explained to the young lady the unfavourable conditions under which he had to work in the wooden workshop at the Chemistry Institute. How spacious, in contrast, was the newly erected building in Dahlem. This Chemistry Research Institute must be thoroughly looked over.

Dahlem was at that time an almost rural area. The newly constructed Institute building stood in the middle of fields which stretched out into a wide plain. It was called the Kaiser Wilhelm Institute of Chemistry and was erected by the newly founded Kaiser Wilhelm Society.

Hahn radiantly told his escort that he was going to work there in the near future as the leader of a section, therefore he would have a tenured position.

Thereupon the couple together hatched plans for the future and became engaged in a subsequent walk through the nearby wood. The wedding was to take place on the 22 March 1913. Outside his nearest relatives Otto Hahn had invited to Stettin a few colleagues who were good friends. Lise Meitner was not present. But she, also, soon formed a warm relationship with the wife of her colleague.

6

The Kaiser Wilhelm Society

6.1   A 'Call to the Nation'

In October 1910 Berlin University was festively decorated for the occasion of the centenary of its foundation. In the ceremony on October 11 in the new Great Hall the Kaiser wanted personally to take to the lectern. Wilhelm II seldom missed such an opportunity to present himself to his people as their true sovereign lord, and on this occasion as the 'great friend and patron of the sciences'.

Many who secretly hoped that this jubilee might lead to a rallying of science and progress were disappointed. Karl Liebknecht, an eyewitness of the celebrations, spoke in the Prussian parliament of 'Byzantine kitsch' and declared, "The entire celebration has led to a glorification of the Hohenzollern generation which would have praised it as the source of all scientific wisdom and achievements".

On the occasion of the centenary celebration the Kaiser announced in a 'call to the nation' the foundation of a society for the promotion of science, in view of which it would bear his hame. In fact, as Wilhelm II, he had already made donations of almost ten million marks which 'from selfless motives' commerce and industry had raised.

There were real grounds for the intended concentration of scientific potential. Under the conditions of the commercial potentials of developments in Germany at the beginning of the twentieth century, science was allocated a specific function. The desired technical prerequisites for profitable production were no longer resulting from the teaching activities of the universities and schools. Ideas would proliferate only from scientists freed from the obligations and duties of a school, in special research institutes built for them. Certain directions in research were difficult to build up in the usual university framework. Organisational aid was permanently complicated and intertwined with an ever increasing budget, which exceeded the means of most universities. A specialisation in research, tied to the concentration of the financial resources in particular institutes, should result in effective scientific work—especially the blooming chemical industry pressed for the expansion and concentration of a separate research capacity. Important scientists like Emil Fischer, Walter Nernst, and Carl Duisberg made themselves advocates of such a development.

Leading industrial countries already possessed state or privately funded research operations from the performances of which commerce profited. Examples of these are the Royal Institution and Lord Ramsay's research institution in England, the Collège de France in France, the Nobel Institute under Arrhenius in Sweden, and the Carnegie Institutes in the United States.

In the search for a suitable person who could present the Kaiser with a memorandum on the foundation of the scientific research centre, the relevant Minister recommended the theologian Adolf von Harnack. As the General Director of the Prussian State Library von Harnack had experience in the field of scientific organisation. Supported by Emil Fischer, Adolf von Harnack presented a secret memorandum on 21 November 1909. Couched in the customary official style of the day, the text begins as follows, "Your Imperial and Royal Majesty has through tireless care for the progress of science and the need for research institutes devoted his most gracious interest ... ".

Hymns of praise to Wilhelm II were to be replaced by sober facts in the subsequent text. Adolf von Harnack did not shrink from voicing uncomfortable truths; German science, before all the disciplines of natural science, was in serious difficulties. "Our leadership in the area of natural research is not only endangered, but we have already had to hand it over to foreign countries in important areas. This hard fact is now already disastrous from a national political standpoint, and it will become ever more so economically." In order not to be further outstripped by other industrial states, close cooperation between the state and the private capital market was suggested.

In his memorandum Harnack put forward to the Kaiser the proposal, on the occasion of the centenary of Berlin University, "to found a Kaiser Wilhelm Institute for research in the natural sciences, for which a suitable plot of land in Dahlem was the most favourable to mention, and to found a chemical research institute as the first institution". Harnack's accounts "met with the most lively and unreserved approval of His Majesty".

Up to the founding of the Kaiser Wilhelm Society in January 1911, 150 firms, bankers, and members of the nobility and squirearchy had raised 10.4

million marks. Around eight million marks were donated by fifty financial magnates. The largest sums given were by Gustav von Brüning, General Director of Hoechst Farbwerk, Krupp von Bohlen und Halbach of the Rhein–Westphalia heavy industry, the bankers Leopold Koppel, Eduard Arnhold, and Franz von Mendelssohn of Berlin, as well as Rothschild heirs. 10.4 million marks for the development of the sciences—certainly in the Wilhelmian era at its height a hitherto unknown budget. Just how should this sum be assessed, since it has been admitted that the Kaiser's Germany had spent 1,300 million marks every year on the fleet and army?

The foundation day of the Kaiser Wilhelm Society was to be 11 January 1911. The number of guests to the constituting assembly was shrunk to eighty nine chosen people. There were only four scientists on it, Adolf von Harnack and Paul Ehrlich, both without a right to vote, as well as Carl Dietrich Harries and Ernst Darmstaetter as benefactors. The imperial couple appeared in order to carry out the act of foundation. At the subsequent official function the first German Nobel Prize winner, Emil Fischer, gave an experimental presentation on 'New Successes and Problems of Chemistry'. He spoke about the manufacture of ammonia from fresh air, about fats and carbohydrates which could be manufactured in the laboratory, about artificial dyes and synthetic rubber and referred to the productivity of industry, the recently produced pharmaceuticals like caffeine, Veronal (barbital), and Salvarsan (arsphenamine). At the end of his lecture he introduced the 'very considerable assembly' to radium research, a branch of science pregnant with the future.

"In this most important investigation Germany at first took only a small part, although the stimulus for the discovery of radioactivity started out from Röntgen rays", went on Fischer. "This lack was especially noticeable because radium also found various useful applications in medicine. It is good to know of its recent production by Privatdozent Professor Otto Hahn. He has busied himself for a number of years with the production of thorium, which is needed in large quantities for the manufacture of gas mantles. At the same time he has discovered several radioelements and has named the most important mesothorium.... I am in the situation of being able to show you a sample of the Hahn preparation. Regarding the strength of radiation, this preparation corresponds to one hundred milligrammes of pure radium bromide, but has cost one third as much. Nevertheless it is not cheap, as eleven thousand marks was paid for this small quantity.... By means of this invention the shortage of radium which has hitherto existed in Germany has now been eliminated.

Twelve days after the foundation festivities the Senate was constituted. Twenty senators were "named by the will of the Kaiser". The choice of Adolf von Harnack, an academic, as President seemed necessary, in order to make the character of the Society correct. As Vice-Presidents two business experts were chosen, Krupp von Bohlen und Halbach and the banker Ludwig Delbrück. Franz von Mendelssohn was to be Treasurer of the Society.

Emil Fischer was to act as Secretary. Under the twenty senators there were four scientists beneath Adolf von Harnack, who as president also belonged to the senate. The most weight lay on the side of representatives of heavy industry and banking. Well known names were there: Gustav von Brüning, Fürst Henckel von Donnersmarck, Krupp von Bohlen und Halbach, Arnold von Siemens, and Privy Commercial Councillor[1] Leopold Koppel.

Many first rank academics felt themselves informed of the advantageous possibilities of working at the institute of the new society. It soon appeared that the basic research at the research locations of the Kaiser Wilhelm Society experienced an exemplary upswing. This fact had a very propitious effect for further scientific progress. Significant discoveries and inventions which were made in the ensuing period proved it only too well.

"The Kaiser Wilhelm Society stands under the patronage of His Majesty the German Emperor, the King of Prussia", asserted the statutes, and Wilhelm II gave it to be known that he thought to perform his patronage and his promised support. "In order to give a sure token of My recognition and good will, I hereby give the members of the Society the right to wear in their button hole a ribbon with My portrait as an emblem woven in orange".

On such gracious gestures the Society of course could not exist. In spite of the start up capital, money was soon lacking. After the Prussian Finance Minister had categorically declined each donation, the existence of the Society was already looking as though it was endangered with a short life span. In despair the President, Adolf von Harnack, directed a letter on 20 November 1911, to the Kaiser and outlined therein one further time the "great task which the Prussian state in our time is engaged in, namely concerning the foundation of research institutes that would fall behind no other nation, and thereby keep at the pinnacle of the sciences". He recalled the traditions of the Prussian state and finally met with a response, certainly because of the following statement, "Military power and science are the two strong pillars of Greater Germany".

6.2   The Kaiser Comes!

In October 1910, a short time after the University's celebrations, Privy Councillor Fischer held a discussion with his radiochemist. He asked Otto Hahn whether he might not be interested in directing an independent radium section in the soon completed Kaiser Wilhelm Institute for Chemistry. What an enticing offer. Hahn immediately decided upon it, for he ever more sharply sensed the inadequacies of his working conditions in the wooden workshop. Fischer promised him he would take into account his wishes about the construction and equipping of the Institute.

---

[1]See Note 8 in Translator's Notes on the Text.

As Fischer explained, the Kaiser Wilhelm Institute for Chemistry in Dahlem would be composed as follows: one division for inorganic chemistry under the direction of Ernst Beckmann, who would also be the Director of the entire Institute; one division for organic chemistry under the direction of Richard Willstätter, specially summoned from Zürich; and a small division for radioactive research.

The first two research locations of the Society, the Kaiser Wilhelm Institute for Chemistry and the neighbouring Kaiser Wilhelm Institute for Physical Chemistry and Electrochemistry with Fritz Haber as Director, would be ceremonially opened on 23 October 1912. The Patron of the Society announced his presence at this occasion. That drove the organisers to a special order of celebration which the participants prescribed "after the end of the ceremony to linger awhile in the library and then to makes one's way outside, by special invitation, making use of the adjoining staircase".

The Kaiser is coming! The programme was rehearsed in all its particulars. Otto Hahn shrank like a delicate rose closing. He was to demonstrate a 'radium illumination effect' to Wilhelm II and his retinue. An Aide-de-Camp of the Kaiser reconnoitred the terrain the day before and had all the experiments explained to him. Safety was the supreme requirement. Real chemists could produce bombs and poison gas without any difficulty ... .

Decoratively laid upon a velvet cushion, 300 milligrammes of Hahn's mesothorium was to be presented to the Kaiser. No value was attached to lead partitions for the shielding of the radiation. In addition it was planned to show the gaseous emanation of radio-thorium, which moved like a ghostly veil over a light screen. Admittedly these attractive experiments had their negative aspects. The fluorescing mist of the emanation and the shinings of the mesothorium could only be made out in the dark. Hahn sought to make the Kaiser's Aide-de-Camp understand. But he remained mistrustful.

"That is out of the question. We can not possibly send His Majesty into a completely dark room".

Long discussions with the quickly summoned Emil Fischer finally ended with a compromise. A small red lamp would light the path of the Ruler.

On 23 October it rained incessantly. Kaiser Wilhelm II, in uniform and with sabre in tow, proceeded with his retinue through an inquisitive guard of honour to the newly built Kaiser Wilhelm Institute for Chemistry. At the main entrance he was greeted by the Minister, von Trott zu Solz, the President, von Harnack, and Emil Fischer. On the second floor the Institute Directors Beckmann and Haber were already waiting, as well as their Divisional Directors Willstätter, Hahn, and Gerhard Just.

Emil Fischer made the speech of welcome and delivered some notable words: "Only he who approaches the great wonders of nature with a simple and modest way of thinking may hope to solve its puzzles by profound and tenacious work. For that the places of experimental research should be free of any sumptuousness, but fitted out with every aid of continually advancing technology." From now on one should "expect dazzling discoveries and profitable inventions from science, fame for the Institute, and benefit and honour to the Fatherland".

During the perambulation through the Institute's rooms, Beckmann the Director explained his methods for determining the molecular weights of chemical compounds. Professor Willstätter demonstrated the isolation of deep green leaf colouring chlorophyll. Hahn's radium illumination effect worked excellently. His Majesty had *not the slightest inhibition about going in the darkened room*, remarked Hahn. Lise Meitner was at first content to remain in the background, but was finally presented to the Kaiser, who affably exchanged a few words with her. For the afternoon the chosen ones expected a special mark of favour from the Kaiser, an invitation to tea in the new palace at Potsdam–Sanssouci.

6.3   Radiation Protection—At That Time a Foreign Word

"How strong was the radioactive preparation that you presented to the Kaiser?"

So nothing was asked about the safety of Wilhelm II by the official concerned at the time. Whatever for? Regulations for dealing with radioactive substances did not then exist. One had no idea of the dangers of such a substance. Nowadays there are strict radiation protection regulations, and Otto Hahn was introduced in an interview in 1967 with that very question.

There were 300 milligrammes of pure mesothorium. I showed it to the Kaiser on a tray, Professor Hahn answered.

"Without any precautions, without protection against the strong radiation?"

Yes. If I had done that today I would be sent to prison. But in those days . . . .

"Did you also work in the laboratory without any protection?"

We always picked up our preparations with our hands; we touched inside them. Under the table on which Lise Meitner and I worked there was a box which permanently contained 150 to 250 kilogrammes of uranium salt. Chemists and physicists would cross themselves[2] today if they had to be radiated upon every day by 150 kilogrammes of uranium salt. I occasionally had sore fingers. But it passed away. Only the nail of my left index finger no longer grows. But of serious impairments I can report nothing.

Because of its dangerousness, the limit for the frequent contact with radioactive substances in the laboratory is an average of 0.1 to 20 microcuries. 300 milligrammes of mesothorium represented, however, an activity of 0.3 curie, that is to say, 300,000 microcuries. Measured against

---

[2] *Translator's Note:* That is to say, to make the *sign* of the cross.

today's regulations Otto Hahn ought only to have handled preparations with an activity one hundred thousand times weaker.

A colleague, the American physicist Ernest O. Lawrence, once said to Hahn, shaking his head, as the latter explained his 'sins' against radiation protection, "It is an impudence of you still to live".

Lise Meitner was in no way inferior to her colleague. During a round in 1959 of the officially opened Hahn–Meitner Institute for Nuclear Research in Berlin, fitted out with all the refinements of radiation protection, she said quite seriously to the eighty year old Hahn, "But at any rate we regularly washed our hands on account of alpha contamination". Hahn nodded thoughtfully and rejoined without taking the cigar from the corner of his mouth, *Yes, and I got so used to doing it that I still do it today.*

Otto Hahn once remarked that he would not have been able to make his great discoveries at all nowadays because of the need for strict compliance with the obligatory regulations, whether the finding of radio-thorium, the enrichment of mesothorium, or the fission of uranium. On the other hand, he knew, however, that radium researchers like William Ramsay, Marie Curie, and the Joliot-Curie husband and wife team, had to pay for their barely tamed urge for research with long drawn out illnesses, infirmity, and even with their lives.

6.4   Research Work at the Kaiser Wilhelm Institute

In his life's recollections Hahn spoke about how he had had to view the foundation of the Kaiser Wilhelm Society as *a quite extraordinary stroke of luck. In the Kaiser Wilhelm Institute now, in spite of modest budgets, everything was certainly much better, and there were five very fine laboratory rooms, a large chemical laboratory and a physical measurements room.* The work spaces were free of the feared radioactive contamination which had already made the physical measurements difficult if not impossible. In addition there was a dedicated workshop with mechanics at one's disposal.

In the spring of 1913 Otto Hahn, together with Lise Meitner, moved into a part of the ground floor room in the right wing of the Institute building. On the first floor was the laboratory of Professor Willstätter, and on the top floor the work room of the Director, Professor Beckmann, and the library.

Beckmann, restricted in his research by the obligations of an institute's director, experimented with the refinement of the method of fixing molecular weights named after him, and devoted himself to molecular rearrangements in organic chemistry.

Professor Willstätter, who had drawn in tried and tested collaborators to the Institute, carried on his investigations into leaf and flower colourings. The results of his research soon found interest, especially amongst the ladies of the Institutes and the many residents of Dahlem. In fact, Willstätter had had a large empty field in front of the Institute stocked with flowering plants. Almost the whole year there waved between Thielallee and Faradayweg, where was the neighbouring institute under Haber, a magnificent sea of flowers. From time to time the splendour of the flowers moved into a stone tub, was crushed, ground, extracted, and chemically separated. For his work on the elucidation of the constitution and on the synthesis of the colouring agents of flowers, Willstätter was, in 1915, the first scientist of the Kaiser Wilhelm Society to receive the Nobel Prize. In 1916 he left Berlin. He took up a call to the University of Münich to follow in the footsteps of his teacher Adolf von Baeyer.

From 14 June 1912, in accordance with the contract, Otto Hahn would for the time being be employed for a period of five years. In the Kaiser Wilhelm Institute for Chemistry he was—according to the text of the contract of employment—to exercise himself "in the capacity of an independent researcher". In addition, he was granted a very meagre budget of 2,000 marks for obtaining materials and paying assistants.

Lise Meitner kept her unpaid position as Guest of the Institute for only a short time. After she had turned down offers from a few universities she was permanently employed in 1914 as a scientific Member. Max Planck had already become aware of her exceptional abilities, and in the autumn of 1912 had provided her with an Assistant's position in his Institute for Theoretical Physics.

Lise Meitner had been one of the first lady Assistants at a Prussian university. She was able to embark upon her habilitation after the World War, when women were allowed to pursue university careers. For that reason this highly gifted woman physicist, whom Albert Einstein liked to call "our Madame Curie", had to wait until 1922 for her habilitation. The number of her scientific publications had grown to fifty five by then. As her habilitation thesis the Faculty of Philosophy of the University of Berlin accepted her work 'On the Origin of the Beta Ray Spectra of Radio-Active Substances'[3]. Her inaugural lecture on 'The Significance of Radio-Activity for Cosmic Processes'[4] even aroused the curiosity of the press. Also it was an unusual occurrence in the Weimar Republic for a woman to hold an authority to teach physics. A printer's error caused involuntary comedy. Thus there was found in a daily paper that a woman had gained her habilitation at the University of Berlin on the theme of 'The Significance of Radio-Activity for Cosmetic Processes'.

When Lise Meitner and Otto Hahn began their collaborative work at the Kaiser Wilhelm Institute for Chemistry, they first busied themselves with completing the radioactive decay series. Unidentified conversion products were to be tracked down and their chemical and radioactive properties studied. Thanks to earlier work, the two scientists were able to order all

---

[3] *'Über die Entstehung der Betastrahlspektren radioaktiver Substanzen'.*
[4] *'Bedeutung der Radioaktivität für kosmische Prozesse'.*

the radioelements known up to then and consequently possessed a 'treasure' which no other research work had up to that time.

Let us pick out just one of the great multitude of scientific problems which interested the pair, actinium. This radioactive element appeared as the continual companion of uranium in all uranium minerals. But actinium is not a long lived element. With a half life of 13.5 years it should not actually have existed longer than the Earth if it were not to have been formed constantly from an unknown long lived substance. Eventually to track down the progenitor of actinium was Lise Meitner's and Otto Hahn's intention. They had already found a name for this unknown element which so doggedly escaped the clutches of their investigations: 'abracadabra'. After a painstaking search the two researchers were still far off the track of the new radioelement when the outbreak of the First World War brought an abrupt end to their work.

FIGURE 6.1. *Above left:* Otto Hahn, 1901, Doctor of Philosophy. *Above right:* Theodor Zincke, Hahn's first academic teacher. *Below left:* The first scientific publication, 1903. *Below right:* William Ramsay, Hahn's second academic teacher, 1907.

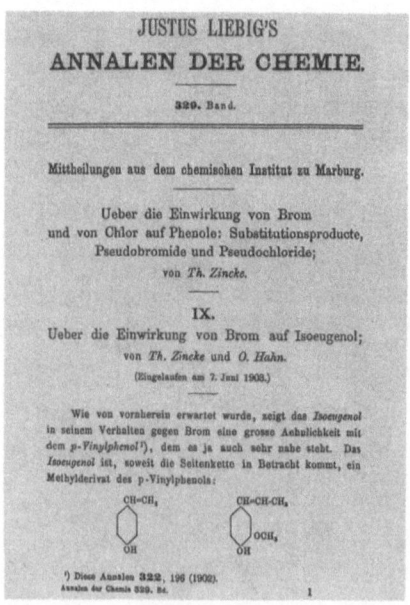 

A New Element.

Very soon the scientific papers will be all agog with a new discovery which has been added to the many brilliant triumphs of Gower-street. Dr. Otto Hahn, who is working at University College, has discovered a new radio-active element, extracted from a mineral from Ceylon, named Thorianite, and possibly, it is conjectured, the substance which renders thorium radio-active. Its activity is at least 250,000 times as great as that of thorium, weight for weight. It gives off a gas (generally called an emanation), identical with the radio-active emanation from thorium. Another theory of deep interest is that it is the possible source of a radio-active element possibly stronger in radio-activity than radium itself, and capable of producing all the curious effects which are known of radium up to the present. The discoverer read a paper on the subject to the Royal Society last week, and this should rank, when published, among the most original of recent contributions to scientific literature.

FIGURE 6.2. *Above:* The Daily Telegraph, 8 March 1905. *Below:* Hahn's third academic mentor, Ernest Rutherford, with his alpha ray apparatus.

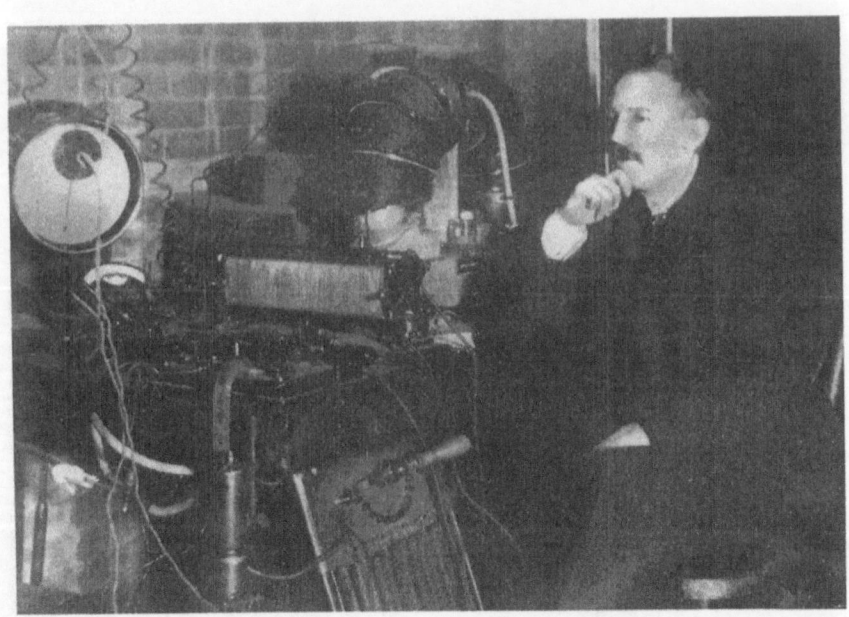

FIGURE 6.3. *Above:* Otto Hahn, Bertram B. Boltwood, and Ernest Rutherford, 1910. *Below left:* Rutherford and Hahn publish jointly the results of their research. *Below right:* Emil Fischer, Hahn's fourth academic teacher.

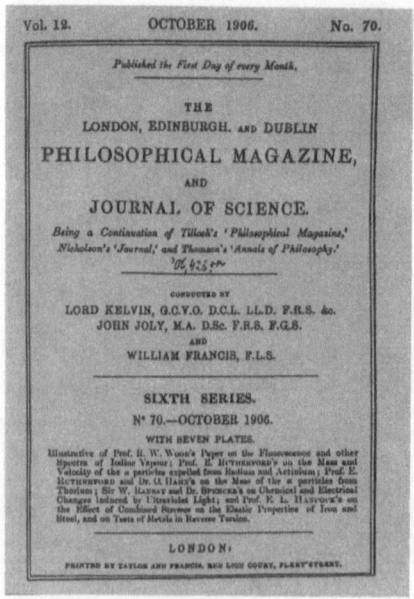

FIGURE 6.4. *Above:* 'An agreeable scientific dispute'. Otto Hahn's greeting from the 1907 Bunsen Congress. *Below:* Leaflet from the Knöfler firm: mesothorium as a substitute for radium.

Dr. O. Knöfler & Co.
Chemische Fabrik.

Plötzensee bei Berlin, Oktober 1911.

Ersatz für Radium!

Seit einiger Zeit wird von uns das von Professor Hahn entdeckte neue radioaktive Element

Mesothorium

technisch hergestellt. Wir bringen damit ein Produkt auf den Markt, das berufen ist, in vielen Fällen das Radium zu ersetzen und vor diesem noch den Vorzug des beträchtlich niedrigeren Preises besitzt. Über seine Eigenschaften orientiert eine Abhandlung von Professor Hahn „Über die Eigenschaften des technisch hergestellten Mesothoriums und seine Dosierung", Chemiker-Zeitung 1911, Jahrgang 35, Seite 845/46, aus der wir die wesentlichsten Punkte in folgendem mitteilen:

FIGURE 6.5. *Above:* The opening of the Kaiser Wilhelm Institute for Chemistry, 23 October 1912. Wilhelm II, with Adolf von Harnack and Emil Fischer behind. *Below:* The Kaiser Wilhelm Institute for Chemistry, Berlin–Dahlem.

FIGURE 6.6. *Above left:* Edith Junghans, later Otto Hahn's wife. *Above right:* The mutual joy of discovery. Otto Hahn and Lise Meitner in their laboratory. *Below:* The mysterious element 'abracadabra'. Lise Meitner's letter of 24 August 1917 (extract) to Otto Hahn, participant in the World War.

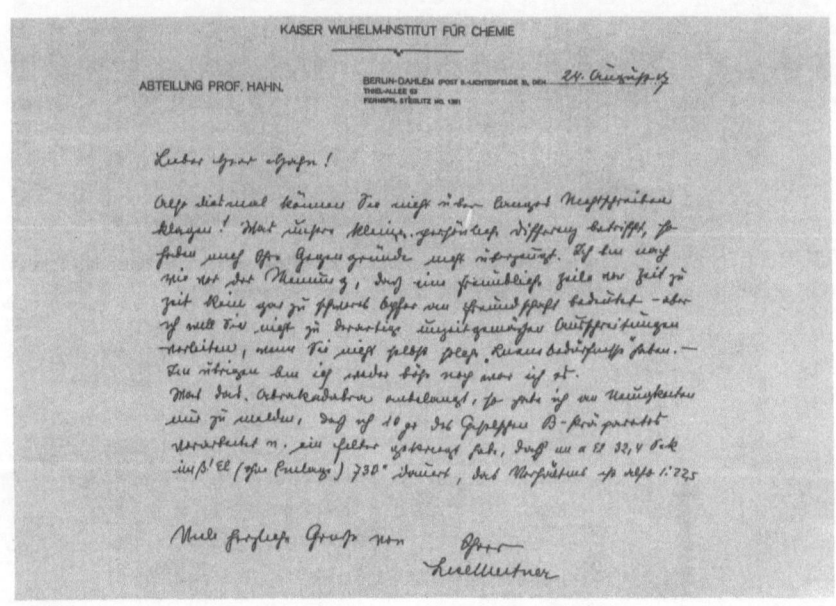

DIE NATURWISSENSCHAFTEN

19. Jahrgang   28. August 1931   Heft 35

Lord Rutherford zum sechzigsten Geburtstag.

Am 30. August 1931 wird ERNEST RUTHERFORD sechzig Jahre alt. Auf der Höhe seines Schaffens stehend, hoch geehrt von den Fachgenossen, bewundert und geliebt von der großen Schar seiner Schüler, zu denen unmittelbar oder mittelbar jeder auf dem Gebiet der Radioaktivität Arbeitende gehört, werden ihm an diesem Tage Beweise der Anerkennung aus der ganzen wissenschaftlichen Welt in reichstem Maße zugehen. Auch wir, die wir das Wachsen und Ausreifen seiner großen Leistungen sehr weitgehend miterleben durften, möchten unsere dankbare Verehrung bei dieser Gelegenheit zum Ausdruck bringen.

Welch eine Fülle glänzender Entdeckungen und orginaler Schöpfungen bezeichnet RUTHERFORDS Lebensweg! Wollte man sie eingehend darstellen, so müßte man die ganze Entwicklung der Physik und der modernen Chemie in den letzten dreißig Jahren schildern. Hier seien nur einige Höhepunkte in Erinnerung gebracht: die Aufstellung der radioaktiven Zerfallstheorie, der Nachweis, daß die α-Strahlen doppelt ionisierte Heliumatome sind, die Begründung der modernen Atomphysik durch die Einführung des Kern-Atommodells, die künstliche Atomzertrümmerung und die durch die Atomkernfelder bedingte anormale Streuung der α-Strahlen. Es gibt kein Problem der Atomkernprozesse, das nicht durch RUTHERFORDS Forschungen maßgebend gefördert worden wäre; aber seine besondere Liebe galt und gilt, wie er selbst vor wenigen Jahren sagte, den α-Strahlen.

Jeder, der das Glück gehabt hat, direkt unter oder mit ihm arbeiten zu können, hat den bezaubernden Einfluß seines hinreißenden Temperaments mit Freude erlebt und sozusagen etwas davon auf sich selbst überspringen gefühlt.

Professor RUTHERFORD, dem die höchsten wissenschaftlichen Auszeichnungen und Ehrungen zuteil geworden sind, hat sich selbst das schönste Denkmal in dem Geiste und der Dankbarkeit seiner Schüler gesetzt.

OTTO HAHN und LISE MEITNER.

FIGURE 6.7. Ernest Rutherford becomes sixty. An appreciation from the pens of Otto Hahn and Lise Meitner (see the translation on page 79).

Lord Rutherford on His Sixtieth Birthday

On 30 August 1931 Ernest Rutherford will be sixty years old. Standing at the height of his creativity, honoured by colleagues in his subject, admired and loved by the crowd of his pupils, who each belong directly or indirectly to the field of work in radioactivity will offer him on this day their evidences of appreciation from the entire scientific world in rich measure. We also, who ourselves were able to witness the waxing and ripening of his great achievements, must offer our grateful admiration through this opportunity to give expression to it.

What a wealth of brilliant discoveries and original creation mark Rutherford's life's work! If one wanted to describe them in detail one would have to describe the entire development of physics and modern chemistry in the last thirty years. Here are just a few high points to recall: setting up the theory of radioactive decay; the proof that α-rays are doubly ionised helium atoms; the founding of modern atomic physics by the introduction of the nuclear model of the atom; the artificial smashing of atoms and through which the field of the atomic nucleus postulates abnormal scattering of α-rays. There is no problem of atomic nuclear processes that was not to be considerably assisted by Rutherford's researches; but his special love was and is, as he has said for some years, α-rays.

Everyone who has been lucky enough to be able to work directly under or with him has experienced the joy of the wonderful influence of his enchanting temperament and, so to say, something of which oneself feels welling up within.

Professor Rutherford, who has been accorded the highest scientific honours and awards, has himself given birth to the most beautiful memorial in the spirit and gratitude of his pupils.

FIGURE 6.8. Participants of the Bunsen Meeting of 1932. *Left to right:* Otto Hahn, Georg von Hevesy, Ernest Rutherford, Hans Geiger, James Chadwick, Lise Meitner, Karl Przibram, Stefan Meyer, and Friedrich Paneth.
Below: Scientific prophecies—for the near or distant future? (A press cutting of 1931). Translation on p. 81.

Das Atom, die Kraftquelle der Zukunft?

Von

Professor Dr. Otto Hahn,
Direktor des Kaiser-Wilhelm-Instituts für
Chemie, Berlin-Dahlem

DN. Die Entdeckungsgeschichte der radioaktiven Substanzen ist wohl ein Schulbeispiel dafür, wie sich aus einer unscheinbaren, fast zufälligen Beobachtung durch folgerichtige Weiterforschung eine kaum übersehbare Fülle neuer Erkenntnisse von allergrößter Tragweite entwickeln kann.

Immerhin hat man aber durch diese künstliche Atomzertrümmerung Kenntnis von Energieansammlungen" innerhalb unserer Atomwelt festgestellt, die man sich noch vor kurzem nicht hätte träumen lassen.

Die große Frage ist daher, wird es einmal gelingen, solche Atomzertrümmerungen und die damit freiwerdenden Energien in einem größeren Maßstabe als bisher mit anderen Hilfsmitteln nutzbar zu machen? Verheißungsvolle Ansätze zur Lösung dieses gewaltigen Problems liegen bereits vor.

Wenn wir vorerst auch noch nicht wissen, wie hier die weitere Entwicklung sein wird, so können wir jetzt, kaum 30 Jahre nach der Entdeckung des Radiums, doch hoffen, daß in abermals 30 Jahren auch dieser kühne Traum der Forschung erfüllt sein wird. Schon heute sehen wir, daß die winzigen Mengen radioaktiver Stoffe, die in der festen Erdkruste in Uran- und Thoriummineralien vorkommen, der Chemie und Physik, der Geologie und der praktischen Heilkunde von unschätzbarem Nutzen sind.

The Atom, the Source of Power of the Future?

by
Professor Dr. Otto Hahn
Director of the Kaiser Wilhelm Institute
for Chemistry, Berlin–Dahlem

The history of the discovery of radioactive substances is a good textbook example of how out of an unspectacular, almost accidental, observation there can develop by consistent further researches a virtually incalculable wealth of new knowledge of the greatest possible consequence.

At any rate, through this artificial smashing of the atom knowledge has been established in our atomic world of the harvesting of energy of which only a short time before we could not dream.

The great question is whether we will ever succeed in such a capturing of the atom's energy and in making it useable on a greater scale than hitherto by other means.

Promising beginnings of the solution of this enormous problem have already been published.

Even if for the time being we do not know what further developments there will be, we can, scarcely eighty years after the discovery of radium, at least hope that within another thirty years this audacious dream from research will be fulfilled. Today we already see that minute quantities of radioactive substances, which occur in uranium and thorium minerals in the Earth's crust, are of inestimable benefit to chemistry and physics, geology and practical medicine.

7
The First World War

7.1   The Defence Forces and Science

As many of his colleagues recognised, in Otto Hahn there was nothing of the
true character of war, in which the leading industrial nations wrestled over
a new division of the world. Above all else, Germany, which had come too
late to the division of the world into colonial regions, wanted new sources
of raw materials and market outlets.

*There was hardly anybody who did not believe in a victorious end to a
just war*, as appeared Otto Hahn to be convinced. A just war? Hahn acknowledged the chauvinistic attitude which initially found an echo in wide
parts of the population. Representatives of German cultural and intellectual life were brought in at the outbreak of war for 'the defence of the
fatherland' and supported in a so called 'call to the cultural world', which
was signed by ninety three academics and artists on 11 October 1914, the
aims of the war of the German Imperial Reich. Germany's blame for the
war was denied, the invasion of Belgium justified, and the imposition of a
great struggle for existence spoken of. Under this 'call' we find the names of
outstanding academics like Adolf von Baeyer, Paul Ehrlich, Emil Fischer,
Fritz Haber, Ernst Haeckel, Walter Nernst, Wilhelm Ostwald, Max Planck,
Wilhelm Röntgen, and Richard Willstätter.

Later, as they grasped the scale of the war and its background, a few of
the signatories, ashamed of their signatures, withdrew their endorsement
or admitted their political and human mistakes. Max Planck was one of

them. Only a few scientists, such as Albert Einstein, refused to give their
approval to the 'call' from the outset.

In the very first days of the war Otto Hahn was called up to a territorial
reserve regiment. Experiences at the front depressed him. He had had to
witness the destruction of the Belgian town of Louvain, an inestimable cultural monument of the middle ages, as it sank into rubble and ashes. The
shelling of the town was justified as a preventive measure because snipers
were suspected to be behind its walls. The eyewitness Otto Hahn was deeply
shaken as students, who had registered as volunteers, were senselessly sacrificed. On the scene of war in Flanders they were driven in swarms into
the fire of English machine gun units. These horror filled pictures engraved
themselves on Otto Hahn's memory.

7.2   Poison Gas Warfare

The following year of the war made a deep incision in Otto Hahn's life. In
his first autobiography, appearing in 1962, we find a single sentence, *After I
was transferred, as Lieutenant of the Reich, at the beginning of 1914 to an
active special unit on the orders of Professor Fritz Haber, in 1917 I came
to the staff of this special unit in the General Headquarters.*

Transferred to a special unit. For the time being Hahn would say nothing
more about it. A few years later he thought differently. In his book *Mein
Leben*, published in 1967, he admitted that at the time of the First World
War he was in unusually many places. He wrote with blunt openness about
his experiences, without any attempt at justification or glossing things over.

In the year 1915 Otto Hahn came in touch with a form of war which
is reckoned as the most loathsome, chemical warfare. For the first time in
history scientists and the military worked closely together on the development of a new device in battle. A crucial stimulus for poison gas warfare
originated with Fritz Haber. The military had first tested these methods of
warfare in the 1870–71 campaign against France; the army had equipped
itself for a blitzkrieg. Only when the German advance on the Marne was
stopped and bitter positional warfare developed did the military change
their mind. Professor Haber then found a hearing when he presented the
possibility of a chemical means of attack and offered his services for developing it for them as a means of shortening the war.

In what way was Reserve Army Lieutenant Otto Hahn involved in this
development? In January 1915 a special order was directed to him. He had
immediately to proceed by the most direct route to Brussels to Fritz Haber.
Hahn was surprised. What had his colleague Haber, Director of the Kaiser
Wilhelm Institute for Physical Chemistry and Electrochemistry, to do with
the war?

At that time Hahn could not have known that the Kaiser Wilhelm Society had undertaken research essential to the war effort. Immediately upon

the outbreak of hostilities the research institutes of the Society were put
into the service of the war effort. Beckmann took instructions from the military for communication by secret light signalling, detection of flammable
gases, safety measures against breathing in of injurious gases. Professor
Willstätter's rooms were seized by the aerial photography command of the
Air Force. He himself worked in Haber's institute on the development of a
gas mask filling. Only the rooms of the Hahn–Meitner radioactive division
of the Kaiser Wilhelm Institute remained spared.

"The Haber people treated us like conquered ground", wrote Lise Meitner, horrified, in a letter to Otto Hahn of 16 November 1916. "First of all,
they were to have only the division's floor, then they required the Lieber-
mann one with it, now they want to take the flyers out of the ground floor
and have monopolised our Chemistry Division, the large technical room,
the weighing room, the photographic and optical rooms were demanded for
their use..., there are endless discussions which are very awkward for me.
I wish that I could creep into a hole and see nothing and hear nothing... ".

In Haber's institute, by the end of 1914 an attempt had already been
made at battle management with chemical gas. At the beginning of the war
year 1916 the institute was then completely converted into a testing and
trials location for the entirety of gas protection and battle methods. Haber
took on the technical direction of the chemical division of the Prussian War
Ministry. The number of staff at his institute, which before the outbreak
of war had shrunk to five coworkers, was to grow during the course of the
year 1917 to 1,500 workers. His budget rose to three million marks.

Kaiser Wilhelm II could not complain at this development. In September 1914 His Majesty had already been "moved to bestow" an Imperial
standard upon the members, "in token of his highest gratification at the
pleasing progress of the Kaiser Wilhelm Society".

Fritz Haber, who had his job in the War Ministry in Brussels, made Otto
Hahn acquainted with the strategic situation. The paralysed western front
would have to be broken through with a new type of weapon. Irritant and
poison gases were under discussion as battle devices. Gas was, so Haber
thought, the best suited to ending the war quickly. In the meantime that
was also the persuasion of the high command. Hahn pulled a dismayed
face and attempted to protest referring to international law and the Hague
convention of 1907.

"Someone will smash your head in", Haber retorted. "We are only responsible for the chemistry and the technical implementation. Besides the
French have made a start... ". The latter argument seemed to Hahn not
very cogent. "If you want to know my opinion, Herr Colleague", before
Hahn put up any further objections, "I consider gas warfare to be legiti-
mate, yes, even for humans. It is an expression of our scientific imagination
which is needed in war and during the preparations for war."

Together with his colleagues Gustav Hertz, James Franck, Wilhelm West-phal, and Erwin Madelung, Otto Hahn was assigned to a new Pioneer[1] Regiment, which was trained in the deployment of gas battle substances. His orders read 'release chlorine gas in a favourable wind in the direction of the enemy's trenches'. With the 'Day of Ypres' the gas warfare began on 22 April 1915. On that day the Germans discharged chlorine gas from a sector of the front in a north easterly direction towards the town of Ypres. The cloud of poison gas took the opposition completely by surprise. "I congratulate you on the beautiful success at Ypres", wrote Lise Meitner very naïvely on 25 April to Otto Hahn out in the field. A few months later, as a Röntgen sister on the eastern front, doing her voluntary year's service, she came to know the horrors of the war.

Chlorine gas on its own was no longer sufficient. In the search for effective battle materials phosgene was swooped upon. With other gas pioneers Hahn found himself on the German front in Galicia on 12 June 1915. They had an order to carry out a gas attack with chlorine and phosgene. Here again, the enemy did not suspect an operation with this insidious weapon. Without a shot being fired, the German troops advanced. *I was deeply ashamed and inwardly very moved*, Hahn later confessed. *We had first attacked the Russian soldiers with gas, and when we then saw the poor fellows lying and slowly die, we wanted to make their breathing easier with our rescue equipment, without, however, being able to prevent their death. That made us conscious of how completely terrible the war was.*

The bitterness towards the Germans of the French and English troops attacked with gas increased. A simple discharge from a gas cylinder was soon outdated, because it depended upon the vagaries of the wind. More effective were gas filled shells, which could be fired with batteries of artillery. Larger battle sectors could be made impassable over considerable ranges. New kinds of poison gas—known under the names of Blue Cross, Green Cross, and Yellow Cross—were used in the battlefield. Blue Cross, for example, was a strongly smelling gas which forced the soldiers to tear off their gas masks, so that they were exposed to the other gases unprotected.

In December 1916 Hahn was transferred to the General Headquarters. While the most bitter positional warfare raged at the front, the history of which up to then is known, Hahn received the order to take charge of the production of gas shells. They were mediocrely produced industrially in the von Bayer–Leverkusen chemical factory under the direction of Carl Duisberg. Also the testing of the protective effect of gas masks in situations of dangerous self-experimentation was included in Hahn's duties. He once received medical treatment because of a phosgene poisoning. After the completion of the work Hahn went back to the front.

---

[1] *Translator's note:* 'Pioneers' and 'Sappers' are traditional army names for engineers. The pioneers originally undertook ('civil') engineering work above ground, and the sappers 'sapped' underground. To sap is to tunnel. Sapping under enemy lines to plant and set off large quantities of explosive was one of the tactics notably used in the First World War.

7.3   A Scientist at the Cross Roads

At the beginning of 1918 the situation of the war worsened further for Imperial Germany. The voices raised against the war became louder and more numerous. Strikes and demonstrations against the war increased, the reign of the tsars in Russia came to an end, and with that, *de facto*, the war in the east.

Yet the revolutionary enthusiasm which gripped the broad masses and also led to the overthrow of the monarchy in Germany remained echoless amongst many representatives of science. The fatalistic composure was reflected in Hahn's words: *Haber had already told us in February of 1918 that he no longer had any hope of a victorious ending of the war. However, we moved forward with our efforts and prepared so many gas attacks....* Admittedly Otto Hahn at first had reservations about war with the insidious weapons of chemistry. *But after Privy Councillor Haber had explained to me how everything was,* he said, justifying himself, *I allowed myself to be converted and later continued out of conviction.... The constant dealings with these strong poisonous substances had so deadened us that we had no scruple whatsoever about deployment at the front.*

Phases of discovery and inward contemplation about the senselessness and criminality of the war alternated in Hahn between an enthusiastic endeavour and "true fatherland performance of one's duty". At that time Hahn, later a convinced humanist and pacifist, was not able to relate to the question of the social responsibility of scientists. The beginnings of a critical assessment of the misuse of scientific knowledge in war were displaced by a questionable loyalty to Kaiser and fatherland.

Most inconsistent of all appeared to be the role of the physical chemist Fritz Haber. With his discovery which changed nitrogen in the air into ammonia and from which urgently required fertiliser was produced, he gave into the hands of mankind a means of defeating hunger. His involvement in gas warfare is, on the other hand, incompatible with the ethos of scientists—to use knowledge and discoveries for the well being of man. However, it would be wrong to lay the blame of responsibility for gas warfare on Fritz Haber alone. The true guilty parties are to be sought in the circles which pursued or hoped for profit from the manufacture of the new chemical battle substances, and to achieve their power political goals with this weapon.

7.4   If Only Atoms Would Fly Into Pieces

With the employment of poison gas the war had acquired a frightening dimension. Since the invention of dynamite by the Swede Nobel there had been no other product of the natural sciences which been so useful in the destruction of life. Despite their enormous effectiveness, these battle materials remained in the realm of man's imagination. But in those years a destructive power of enormous magnitude was already being played with intellectually, namely, atomic energy.

"The most powerful explosive we know", it was called in Frederick Soddy's paper 'Matter and Energy' of 1912, "contains hardly one millionth part of the energy which is freed if an atom flies into pieces". Fortunately, so Soddy believed, mankind at the moment "is no more competent" to use atomic energy "than a savage who wants to use a steam engine and has no idea of how to make fire". But what did that mean? "It could take science many years, perhaps even a century, to find this means, yet the prey is already well in sight, and researchers have already started out on countless routes in hot pursuit".

The First World War shook Soddy's belief in progress. He was already dreaming of a paradise on Earth with the help of the miraculous power of the atom, but he had to acknowledge that he was dismayed that another application for the use of atomic energy had not presented itself. "It may be put forward that if one could have seen what the present war was to have been like, then were such an explosive to be discovered the future might, instead, lie in safe keeping".

H.G. Wells, the highly creative English author, had such powers of imagination and shocked his readers in the novel 'The World Set Free', which he dedicated to Soddy, with pictures of the destructive power of an atom bomb; in 1913 he envisaged towns totally destroyed.

Rutherford behaved with pointed restraint when in 1916, in his lectures on 'The Radiation of Radium' at the University of Manchester, he explained, amongst other things, "Naturally, scientists are attracted to the task of seeking a way to make the enormous energy slumbering in radium useful. They hope, however, that this will not be successful as long as men do not learn to live in peace with their neighbours".

8

New Success for the Atomic Researchers

8.1   The Old Dream of the Alchemists

During the war Rutherford had to provide the British Admiralty with reports on the counter-intelligence possibilities of enemy U-boats. Every free minute he spent on his real research work. With Niels Bohr he cultivated a significant exchange of ideas. Their correspondence vividly illustrated the unbroken scientific enthusiasm for their work during that critical time.

"I wish I had you here, so that I could discuss with you the meaning of a few of my results on nuclear collisions", wrote Rutherford on 17 November 1917 in a "strictly confidential" letter to the Danish physicist. "I believe I have some astonishing results. However, it is a difficult and lengthy beginning of obtaining a sure proof of my conclusions. Counting weak scintillations is difficult for old eyes".

It is not difficult to guess that here it is the favourite occupation of Rutherford which is spoken of. In addition, undaunted he bombarded the fortress that was the atom with his alpha rays in the hope of victory one day. "I am seeking to split the atom by this procedure", revealed a further letter to Bohr of 9 December 1917. "In one case the result looks very promising".

With such rousing prospects it is understandable that Rutherford performed his services for the military only reluctantly. When the Admiralty once reprimanded him for unpunctuality he coolly answered, "I was involved in experiments on the artificial splitting of the atom which I had to shut down. If this can be done it will be of much greater importance than your war!"

The first report of Rutherford's new experiment was to be read in the following April of 1919, the factual text of which was almost disappointing, and was published in the June issue of the *Philosophical Magazine*. The atomic physicist called his paper 'The Collision of Alpha Particles with Light Atoms. An Abnormal Effect with Nitrogen', behind which was secluded an exciting scientific discovery.

With his usual persistence Rutherford had bombarded various gaseous elements with alpha rays and measured the distance through which the gas atoms hit rushed. At the same time he used, as usual, a zinc sulphide screen to count the scintillations caused by the atoms which had been struck. Nitrogen atoms struck by alpha rays were flung about nine centimetres.

When Rutherford—accidentally, or perhaps intentionally?—set the scintillation screen at a greater distance, however, he noticed occasional surprising traces of light which must have come from quite different particles. As he proved, it was a question of atomic nuclei of hydrogen with a range of about twenty eight centimetres in air. It was ruled out that any of these tracks were from hydrogen in the apparatus. Where did these hydrogen atomic nuclei come from?

After some pondering, Rutherford found an astonishing explanation. The hydrogen atom must have been split from the nitrogen atom. Further investigations confirmed the correctness of these ideas. "My work on the atom moves on in an elegant way. Each week a few atoms succumb", he wrote to Bohr. In the meanwhile Rutherford settled into Cambridge, where in the same year, 1919 he took over the direction of the famous Cavendish Laboratory as the successor of J.J. Thomson.

Then the Englishman Charles Thomson Wilson succeeded in making the paths of atomic nuclei and other charged particles more visible for human eyes as tracks of condensation in a cloud chamber, and everywhere people found the conversion taking place that Rutherford had reported, that of a fork instead of the otherwise customary simple track.

Rutherford's coworker P.M.S. Blackett took large numbers of photographs of such nuclei tracks in a Wilson cloud chamber. It speaks for the improbably low chance of a strike or conversion that Blackett had to analyse over twenty three thousand photographs in order to find eight pictures with a fork. In these eight cases resided the conversion reported by Rutherford.

A nitrogen atom with atomic weight 14 was converted by an alpha particle (helium) into an oxygen atom of atomic weight 17 and one proton, that is to say, a nucleus of a hydrogen atom.

This was the first time one element had been successfully converted into another artificially. What the alchemists had unsuccessfully sought for centuries, to transmute one element into another, into silver or gold, nuclear physics had now succeeded in doing, even if only at the level of the nucleus.

Rutherford calculated that it would take thousands of years to produce a cubic millimetre of hydrogen gas in this manner. The physicist Carl Friedrich von Weizsäcker later found an apposite comparison, as he thought, the chance of smashing an atomic nucleus in this way was "no better than the likelihood of shooting a single flea in a large, completely dark, railway station building by blindly firing a machine gun in it".

Up to the year 1919 researchers had not succeeded in inducing artificial radioactive decay or even conversion of the atomic nucleus—a prerequisite for the extraction of energy from the atom, which was suspected at that time. Now, for the first time, such a transmutation had been successful! "People were on their guard for an illusion", commented Walter Nernst in a lecture given in 1921, "as if the technical production of the quantities of energy available here was just about to make coal worthless".

Nevertheless, there must be a way of liberating this atomic energy. Nature put it before mankind daily. Those immense sources of energy the Sun and the stars radiate constantly—and that could be nothing else than atomic energy.

A coworker from Rutherford's team, F.W. Aston, who in 1922 constructed a mass spectrometer for the separation of isotopes, then sought to solve the riddle of atomic energy in a theoretical way. Supported by Einstein's mass–energy equation, he came to the conclusion that the atomic nuclei of the heaviest elements, such as radium, thorium, and uranium, were converted into nuclei of moderate mass. When Aston committed the mathematical proof to paper, a feeling of horror overcame him at the sight of the gigantic numerical values: "Should a way ever be found, this occurrence would probably reveal itself as the birth of a new star... ".

The alarming idea that through some kind of detonation in some kind of place atomic energy might be released and communicated to other atoms, and that finally the energy of the entire matter of the globe might be converted into the energy of a radiating star, built its nest firmly in the minds of many academics. Walter Nernst expressed this vision figuratively in 1921: "the existence of mankind as a primitive people which lives on an island made out of gun cotton, but does not have any fire".

It might be thought that this view might have stopped researchers from digging deeper into the secret of the atom. But apparently there was only one question for them, that of how to find the 'match' needed to set alight the atomic fire.

The urge for knowledge, the advance into unknown areas of objective reality, has always been stronger than the fear of the unknown, of which the history of the natural sciences is the proof.

8.2   The Father of the Race, Protoactinium

His temporary detailing to Berlin during the war was not inconvenient for Professor Hahn, as he could take a little care of his radiochemistry research. When he had cause to fear the military would also commandeer his measuring room, he wrote promptly to Lise Meitner. His colleague was at this time doing her voluntary service at the front. At Otto Hahn's urgent plea she went back to Dahlem. According to her own words she felt very unsure of to what extent she might oppose the orders of the new gentlemen of the Institute. In this difficult situation she obtained the advice and help of Max Planck, who energetically applied himself for the continuation of the scientific work.

So it was that the search, begun in 1913, for the progenitor of the element actinium was finished during the war. It confirmed the suspicion that the unknown long lived radioelement would have to be sought in the very last residues of the pitchblende from which the uranium and radium had already been separated. This residue, which was especially difficult to break down, typically called 'the grey misery'[1] in the industry, Hahn and Meitner subjugated with a threefold breakdown with aggressive chemical agents. They succeeded in enrichening the radioelement sought and in determining its constants.

To their delighted surprise the two researchers soon found out that not only had they quite simply discovered a new radioactive isotope of especially long life, but also a completely new chemical element that took up the hitherto unoccupied position 91 in the periodic table. The second heaviest element that existed on Earth had been found. Long jokingly dubbed 'abracadabra', they christened it protoactinium.

Hahn and Meitner called their joint paper 'The Mother Substance of Actinium, a New Radio-Active Element of Long Life-Time'[2], which was submitted to *Physikalische Zeitschrift* on 16 March 1918, and published on 15 May.

With the discovery of protoactinium the last gap in the three series of the natural transformations of the radioactive elements was closed. All representatives had been found. An important area of Otto Hahn's work which he had made his task—the finding of the intermediate radioactive elements—had therein met its conclusion.

Professor Hahn had wanted to devote himself henceforward to problems of applied radiochemistry, when he made a further unexpected discovery which succeeded in explaining a baffling effect. By looking back on the known isotopes UX_1 Hahn noticed a weak beta radiation which came from a real but unknown radioelement. In order to remove any mistake he decided on a painstaking operation, in which he processed one hundred kilogrammes of uranium salt to obtain a larger quantity of UX_1, and noted the decay carefully for one hundred and sixty five days. By this investigation he was able to prove that UX_1 decayed in two ways. Chiefly, as is well known, it decayed into UX_2, but 0.35 percent also decayed into a new

---

[1] *'Das graue Elend'*.

[2] *'Die Muttersubstanz des Actiniums, eines neues radioaktives Element von langer Lebensdauer'*.

radioelement from which the inexplicable remaining radiation came. Hahn called it uranium Z (UZ).

The two substances UX_2 and UZ were not only isotopes, that is to say, belonging to the same chemical element protoactinium (Pa_{234}), but also isomers, as they had the same atomic weight. Their distinction can be made only on the basis of their decay properties. Otto Hahn discovered to his own surprise something further, which he had not been seeking at all. He was the first scientist to report a nuclear isomerism. His research paper 'On a New Radio-Active Decay Product in Uranium'[3] of 21 January 1921 can be read in the issue of *Naturwissenschaft* dated 4 February 1921.

It was not wrong of Hahn later to describe the complicated explanation of nuclear isomerism as *his best work of all* and to add that it had been suspected a little of a Nobel Prize. But the scientific world of 1921 did not honour his discovery, and on a trivial ground; nobody knew what to do with it. Well known physicists had later said that Hahn's discovery had been fifteen years too early—an astonishing judgement in view of the huge rate of discovery in atomic research. It was only in 1936 that the physicist Carl Friedrich von Weizsäcker, then active in Hahn's Institute, first gave an adequate theoretical explanation of nuclear isomerism, and from 1935 German and Soviet researchers had put into order further examples they had demonstrated.

Another 'discrepancy' remained in Otto Hahn's mind with, as usual, the customary consequences. According to the correct picture of actinium there remains a residual activity of one percent. *I can not make up my mind whether to attach any importance to this small quantity*, thought Hahn. A French researcher, Marguerite Perey, followed up this path, admittedly only much later, down which she found the missing element number 87, which she called francium.

8.3   No Luck Without Service

It is also safe to ascribe to the achievements of Otto Hahn and Lise Meitner the rapid return to international recognition of German science after the First World War. At a time when German scientists were still barred from international meetings, the Dane Niels Bohr was the first to break this isolation and invited the physicist Lise Meitner together with James Franck to colloquia at Copenhagen.

When the famous physicist, who received the 1922 Nobel Prize for Physics for his research into the structure of the atom and radiation of atoms, was consequently invited to Germany he gladly took up the invitation. 'Bohr Festivals' soon became a term in Germany. The young generation of

---

[3] *'Über ein neues radioaktives Zerfallsprodukt im Uran'*.

physicists, especially, felt themselves drawn to Bohr's personality. Werner Heisenberg, Max Born, Pascual Jordan, and Erwin Schrödinger profited from Bohr's great wealth of theoretical knowledge and soon enlarged atomic physics with new ideas such as the epochal theory of quantum mechanics. The international family of atomic researchers moved still closer together.

Otto Hahn and Lise Meitner, for a long time employed as 'permanent scientific members' in the Dahlem Kaiser Wilhelm Institute for Chemistry, expanded their division further. Professor Alfred Stock, who had taken over the direction of the institute since the return of Beckmann in 1921, wrote in 1924 a worried comment on the financial plan, to the effect that the Hahn–Meitner division surpassed the budget of the two other divisions of his institute by some measure. The *radioactive contamination* seems to have grabbed a greater and greater hold of the entire Kaiser Wilhelm Institute for Chemistry, as Hahn mischievously said.

Otto Hahn's scientific achievements found increasing recognition. From the Prussian Minister for Science, Art and Education he had the title of a non-permanent Extraordinary Professor conferred upon him. At their meeting of 6 November 1924 the Prussian Academy of Science of Berlin elected him to Ordinary Membership of the physical–mathematical section 'by thirty one white balls to two black'.

In the application for membership, which Haber, Wilhelm Schlenk, von Laue, and Einstein supported, it states, amongst other things, that Otto Hahn owed his place in science to his work in the field of radioactivity, which had led to the discovery of no fewer than nine new radioactive elements.

Indicative of Hahn's modesty is his inaugural speech before the Academy on 2 July 1925, which began with the following words: *If the great honour is shown to me today of being accepted into this circle of important and highly respected academics, and if I seek to clarify the grounds for so being, there steals over me a feeling of unworthiness, for I owe this honour to a series of lucky accidents . . .* . Radioactivity research, Hahn gave to understand, is a young science. He who dedicates himself to it totally therefore has it easy in experiencing the joy of discoveries, which in other branches of science is much harder to come by.

Max Planck, in his capacity as Secretary, replied, "As you yourself have just emphasised, the novelty and virgin territory of the field worked in by you has certainly contributed much to your successes. But if in hindsight you attribute a particularly prominent role to lucky accidents in the development of your research, then I ought to change my opinion that in science sometimes one must give service without luck, but quite certainly never luck without service, even if the connection can not always be demonstrated in an obvious way . . . . Thus we may certainly expect many new discoveries and many new solutions at the fruit of your work".

In the course of his life of assiduous hard work Otto Hahn was to become a member, or, to be precise, an honorary member, of forty five academies and scientific societies.

8.4 Applied Radiochemistry

To Otto Hahn the work in the newly built Kaiser Wilhelm Institute in its first years appeared to be a stroke of luck, so it should have changed during the time of the great inflation. The Kaiser Wilhelm Society, too, was unable to provide the means for maintaining the research work. The number of coworkers shrank very considerably and even the time for work did not escape the effect. Laying off, short time, and unemployment did not bring science to a halt.

When the Director Stock retired from the Institute in 1926, Otto Hahn took over the provisional leadership. To the joy of his coworkers he had turned down a call to the Technische Hochschule of Hannover. As 'Director for Life', Otto Hahn received a contract with conditions advantageous for that time. His yearly salary came to almost 24,000 marks. The consequent improvement in social position was also to be reflected in the quality of life of his family. In March 1929 Edith and Otto Hahn together with their seven year old son Hanno took a house in Berlin–Dahlem. The first prominent guest in their new home on the Altensteinstrasse was Ernest Rutherford at the beginning of May 1929, who was fulfilling commitments to lectures in Berlin.

In scientific questions with now only himself to answer to, Hahn turned the Institute round to his plans. 'Radioactivity' had now actually spread 'like an epidemic' through the Institute's rooms. Only a small guest division for organic chemistry remained in the building. Professor Lise Meitner had advanced to Director of the Physical Radioactivity Division. Otto Hahn still kept the Chemical Radioactivity Laboratory under himself. Even as directors of independent divisions, Hahn and Meitner kept up their tried and tested joint work.

'Applied Radiochemistry'—with this concept one could summarise the field of work which from now on Otto Hahn prescribed for himself. His investigations continued to be characterised by meticulous work with unmeasurable amounts of substances, an operational technique which Hahn had developed to a craft of consummate mastery. From then to today they are proclaimed as the basic knowledge of radiochemistry, known as the 'Hahn precipitation and absorption rules'.

On his instructions, for example, tests were sought for various salt deposits, in order to interpret the inexplicable occurrence of lead and helium in these salts. Hahn and his coworker solved this puzzle; both elements came from the radium of the prehistoric ocean.

An important result of these works in applied radiochemistry is the publication 'What Does Radio-Activity Teach Us about the History of the

Earth?'[4], Otto Hahn's first scientific monograph, which appeared in 1926 from the publishing house of Julius Springer of Berlin.

A number of new experimental methods were developed. With the help of the 'emanation method' worked out by Hahn, which monitored the delivery of the easily demonstrated gaseous radio-emanation, one can in an elegant way, namely through the radioactive measurement of the emanation set free, investigate the surface and structural properties of a particular substance.

Hahn's 'rubidium–strontium method' has become a well known procedure for fixing geological age. Also, there is a very wide variety of uses for this method, for example in the investigation of meteorites or tests on the lunar rocks brought back to Earth. Although this is made use of by a cleverly devised piece of automated equipment—a masterpiece of electronics— such as may be found in any modern nuclear physics or radiochemistry laboratory, in those days there was none such. Karl Erik Zimen, an assistant of Otto Hahn, has emphasised in his 'Memories of the Kaiser Wilhelm Institute for Chemistry'[5] that at that time all equipment was marked by a great simplicity. Hahn's ideal still remained the electroscope made of tin cans for preserves. "After we had strained our eyes with them fixed to the eyepiece of the electroscope for a year", related Zimen, "We lighted upon the idea of building a machine aid". It was a matter of honour for the Institute itself to make the Geiger–Müller counter tube, instead of obtaining it from outside, despite the danger that of five made only one worked. When modern equipment and apparata came into the Institute Otto Hahn at first explained each unexpected 'effect' as the spawn of the new and complicated technology.

In his capacity as the Director of the Institute, Hahn uncompromisingly insisted on repetition and checking of experimentally determined values, and was only satisfied if the results could be reproduced not just once or twice, but five or ten times. This attitude to one's work had at the time driven many doctoral students to the edge of despair. Yet all his students later vouched that this apprenticeship with Hahn had been valuable to them, because they were brought up to be self-critical and self-controlled.

Much of the acknowledged good atmosphere of the Institute resulted not only from Hahn's insistence on strongly disciplined work, but above all on his humane attitude, which was manifested in sympathy and understanding for the day to day worries of his employees. "In this working company there ruled a good spirit and a happy mood, a reflection of Hahn's personality", opined Lise Meitner. Otto Hahn's sense of humour seldom did not have a share in that, being sometimes ironic, never wounding, but always having a rousing effect.

---

[4] *'Was lehrt uns die Radioaktivität über die Geschichte des Erde?'*.
[5] *'Erinnerungen an das Kaiser-Wilhelm-Institut für Chemie'*.

When Assistant Zimen once worriedly thought that the air of the radium building contained a disturbingly large amount of radium emanation, Hahn replied, "Be thankful. Others pay a lot of money to travel to Bad Gastein (a radium spa). You get it here gratis".

Perhaps Otto Hahn had wished, all in all, to build an institute such as he had known and come to appreciate in Canada under Rutherford. Possibly he would have succeeded in that if the political developments growing deep within the life of the institute had not intervened. At any rate, Hahn knew how to make the Institute at Berlin–Dahlem a focal point for trainee students, doctoral students, and foreign researchers, who not least of all also felt drawn by his personality. Scientists from Belgium, Canada, China, Great Britain, Holland, Italy, Sweden, the USSR, and the USA chose the hospitable Institute as a place to deepen and improve their knowledge and skill.

8.5   The Remarkable Year 1932

"We have been able to invite Bohr here yet again for an exchange of scientific views", said Otto Hahn on one occasion to Lise Meitner when they came to speak of the two uncrowned kings of atomic research. "We should try to invite Rutherford to Germany to the next large congress..." .

On the table lay a calendar. Hahn leafed through it and seemed to have found a suitable occasion. "How about the Bunsen Meeting in Münster? I think we shall suggest to the Bunsen Society that the next members' meeting be held on the theme of 'Radio-Activity'".

Rutherford agreed to come. He asked to bring a few of his 'fellows' with him, and announced a promising piece of scientific news. When it was heard the Lord Rutherford of Nelson was going to come—the atomic researcher had not long ago been elevated by the King of England—other researchers also announced their reports. Almost all the scientists who had any kind of status and name in experimental atomic research came to the meeting, which lasted from 16th to 19th May 1932. In fact, the only people missing were Marie Curie and the Joliot-Curies from France, and a few American and Canadian researchers such as Lawrence and Eve.

Let us take a look at the plan of events. Ernest Rutherford gave the opening lecture on the theme 'Memories of the Early Days of Radio-Activity'. His pupil James Chadwick gave a paper on Rutherford's sensational announcement, the discovery of a new elementary particle. The Soviet mineralogist and geochemist Vladimir I. Vernadski familiarised the meeting's participants with the theme of 'Radio-Activity and Problems of Geology'. His fellow countryman Vitaly G. Chlopin, Director of the Leningrad Radium Institute, spoke about 'Radium- and Mesothorium-Containing Stretches of Water'.

Lise Meitner's talk was about the investigation of beta and gamma rays. Otto Hahn made observations on the application of emanation methods. His assistant Fritz Strassmann spoke about the intense emanations from barium–radium salts, Friedrich A. Paneth about isotopy, Georg K. Hevesy about radioactive indicator methods, and Hans Geiger about alpha rays.

As expected, the greatest interest was concentrated upon Rutherford's introductory lecture. To the delight of the host the prominent atomic researcher found words of recognition for the achievements of German science, praising his earlier 'promising pupil' Otto Hahn. Ernest Rutherford left no doubt that atomic research would be quickly developed further. "We are only at the beginning of this direction in research and we ought to expect an enormous expansion in the next five to ten years".

At the same time some period of stagnation was in some respects unavoidable. The hope that by using Rutherford's methods all atoms, little by little, would be converted or smashed with high energy alpha rays was followed by disappointment. Ten years after the first successful experiment there were scarcely more than a dozen elements, and then only the lightest, which had fallen as victims to bombardment. In the heavier elements even alpha particles with the enormous energy of nine million electron volts could not even penetrate the massive atomic nucleus. They were either deflected or thrown back by the high charge without coming into contact with it.

The 'electron volt' is not to be confused with the customary 'volt' which is the unit of voltage. It serves only as a measure of the energy of the minute elementary particles. Only after multiplication by the inconceivable number of atoms and particles, for example contained in one gramme of matter, do these amounts of energy of the microscopic world reach gigantic proportions. In order for an electron to be removed from the atomic shell it requires the small energy of a few electron volts (eV). However, to split asunder a building block of the nucleus of an atom, an energy of a few million electron volts (MeV) is necessary.

Also, with alpha particles of several million electron volts' energy one can not reach this declared goal with the heavy elements. There was thought to have been found a way out by using protons, that is to say nuclei of hydrogen atoms, as projectiles. To do that it was necessary artificially to accelerate these particles to high energies similar to those of alpha particles. But how would one obtain the huge energies needed? Voltages of several million volts would have to be produced and stored for this purpose—a technology that had not yet been mastered. Could this energy perhaps be captured from lightning? In 1927 three Berlin physicists set off on an expedition to the fantastic mountain region of the Lugano Alps in order to capture the electricity of a thunder storm with insulated cables on the 1,700 metre high Monte Generoso. One of the scientists of this risky undertaking paid for it with his life.

In an article in the press of the time, entitled 'The Atom, The Power Source of the Future?'[6], Otto Hahn took up the risky experiment on Monte Generoso. *The great question is, will such atom smashing and consequent release of energy on a greater scale than hitherto be successfully made exploitable by other means? Promising starts on the solution of this enormous problem have already been put forward, ... and we are now able, just thirty years after the discovery of radium, to hope that in another thirty years this bold dream of research will be fulfilled.*

With such dreams coming to fruition in due course, this theme even broke into the budget debate in February 1931, in the German Reichstag. An application for a contribution to the costs of the Kaiser Wilhelm Society was substantiated by Max Planck by saying that in Dahlem they "were engaged in investigations which were concerned with the important problem of the smashing of atoms. This question would, in the future, if our coal deposits ran out, obtain an immense significance, for within atoms slumbered tremendous energy". (Quoted from the written minutes).

In order to smash the atomic nucleus and to be able to set free its energy, one needed for this purpose highly accelerated elementary particles. In Washington American physicists constructed a Tesla transformer for three million volts in 1931, and used it to accelerate electrons to an energy of one million electron volts (1 MeV). One year later Robert van de Graaff built his first belt generator of 1.5 MeV, later named after him, at Princeton University.

Ernest Lawrence and his coworkers at the University of California at Berkeley finally succeeded in making a breakthrough. By a trick Lawrence forced the particles into spiral paths with the aid of a large electromagnet. In this way they could be accelerated to a tremendous speed and collected together into a high energy beam. Medium voltages were sufficient for the purpose. The cyclotron had been invented. With this accelerator one could now achieve radiation intensities which were theoretically equivalent to several kilogrammes of radium. That was an enormous step forward, for one would never have been able to separate such a quantity of pure radium.

At the end of 1931 Lawrence reached 1 MeV with his cyclotron; a year later, with a better model, 2 MeV. Nowadays one reckons the performance of modern particle accelerators in giga-electron volts (GeV), that is to say in billions of electron volts. Compared with the first circular accelerator of the American Lawrence, which may be found in every large laboratory, the highest performance particle accelerators of today, with their frequently kilometres long tracks, are gigantic.

The pupils John Cockroft and Ernest T. Walton of Rutherford were the first, in 1932, to succeed in transmuting the nucleus with accelerated

---

[6] *'Das Atom, die Kraftquelle der Zukunft'.*

protons[7]. The atomic nucleus of lithium, which is the lightest element after hydrogen and helium, changed into helium by being bombarded. In the press it was read that with this 'smashing of lithium' a further step had been taken by mankind towards the subjugation of atomic power. A battleship could cross the wide Atlantic with less than a gramme of lithium as 'fuel'.... . In actuality, the transmutation of the lithium occurred with a tiny, immeasurable yield. Millions of protons had to be accelerated in order to achieve even a single collision.

To the tried and tested projectiles which were to be fired at atomic nuclei—alpha particles (He atomic nuclei) and protons (H atomic nuclei)—was added another, the deuteron, which was the nucleus of heavy hydrogen with twice the mass of a proton.

In the same year the positron was discovered in cosmic rays, being the positively charged counterpart of the negatively charged electron. Lise Meitner and her coworkers were the first to succeed in demonstrating the existence of positrons, but by terrestrial radioactive processes, with the help of a cloud chamber.

At the Bunsen Meeting in Münster, in May 1932, Rutherford and Chadwick reported on a further newly found particle from which one might expect fundamental effects—the neutron. James Chadwick had succeeded in this discovery after his teacher Rutherford had predicted the existence of the neutron. As other researchers had also been on the track of the neutron, a revealing history had entwined itself about the discovery of this elementary particle.

In a series of lectures before The Royal Society in the spring of 1920, the Bakerian Lectures, Rutherford had astounded his hearers with a series of astonishing predictions. He was convinced that a nucleus with atomic number 2 and charge 1, which hydrogen had, must exist—the deuteron, discovered twelve years later. In addition, he had given reasons for his hypothesis of the existence of a 'charge-free' proton with atomic number 1 and nuclear charge 0, as this was indispensable for the building of especially heavy atomic nuclei.

But how was one to get on the track of this elementary particle? After constant endeavours to transmute further elements by bombardment with alpha particles, in 1924 Rutherford and Chadwick found an annihilation effect with beryllium like that of nitrogen previously. Both researchers thought, however, that contaminants could have simulated this result.

Six years later the German physicists Walther Bothe and Herbert Becker discovered that the light elements, and above all beryllium, gave off a penetrating radiation if they were exposed to alpha rays. With a cleverly devised method the two researchers were able to demonstrate that it was a hard gamma radiation that arose from the atomic nucleus. Irène Joliot-Curie

---

[7]See Note 9 in Translator's Notes on the Text.

and her husband Frédéric had already experimented for some time with this new beryllium radiation in their Paris radium laboratory. Finally they had unexpectedly established that H-nuclei, and thus protons, had been liberated from beryllium irradiated by these rays. Such an effect, however, was not reconcilable with the then current ideas about the effects of gamma rays. In their work published in the *Comptes Rendus* of 22 February 1932 the two French scientists did not hold back their astonishment. In further investigations they endeavoured to produce a clarification.

They did not have long to wait. Five days later the English journal *Nature* carried a rush announcement from the Cavendish Laboratory, 'On the Possible Existence of a Neutron'. Chadwick had found by checking the investigation used in Paris that the beryllium radiation was not gamma radiation, but very probably consisted of particles of mass 1 and charge 0. But Chadwick's detailed publication followed in June 1932, in which he first spoke of the 'Existence of a Neutron' secured in the interim.

Chadwick's interpretation had not been completely right. Bothe and Becker had reported nothing that was wrong. With their method of measurement they could test for a nuclear gamma ray unambiguously. Chadwick was thus incorrect in his claim that the supposed gamma radiation was actually neutrons, because his apparatus could only identify neutrons and not gamma quanta. In reality, there existed both kinds of radiation.

Owing to the surprisingly speedily successful explanation of the beryllium radiation, Chadwick's presentation, although at first not planned, was included in the programme of the Bunsen Meeting at short notice. Walther Bothe expressed his admiration: "After the beautiful investigations of Chadwick no doubt can arise about the existence of neutrons".

The discovery of neutrons had nearly been made in the Kaiser Wilhelm Institute for Chemistry. Lise Meitner had, as well, predicted the existence of such a building block of the nucleus theoretically in 1921, but was too late to think of an experimental check. She first encouraged a foreign guest of her division to do so, the Italian Franco Rasetti. Under Lise Meitner's instructions, Rasetti was able to explain the complex nature of the ominous beryllium radiation: gamma rays together with neutrons. His work was, however, ready only in manuscript on 15 March 1932, and appeared in *Naturwissenschaft* of 1 April. In the meanwhile, *Nature*'s issue with Chadwick's results had already been published.

The wed Joliot-Curie researchers were also disappointed. The discovery of the neutron had only been narrowly missed in the Paris radium laboratory. Both scientists were convinced that they succeeded in finding the neutron a long time before Chadwick if they had known about Rutherford's prediction of the chargeless particle. This time those in Cambridge had been quicker.

When in 1932 the neutron had been discovered, that year was rightly called in specialist circles an 'annus mirabilis', a wonderful year.

With the finding of the particle each difficulty which the theoreticians had encountered in the build of the atomic nucleus was instantly solved.

Finally the opinion was that the atom consisted of a nucleus of protons whose positive charge was compensated by that of the circling electrons. Such an idea, however, salvaged insoluble contradictions for the time being. Likewise, they delivered no explanation of why with identically charged nuclei the isotopes of an element were quite different. Werner Heisenberg, and independently of him the Soviet physicist D.D. Ivanenko, discovered a new model of the structure of the atomic nucleus in which one had to think of it as made up of protons and neutrons together. The various masses of the isotopes were then explained by an excess or deficiency of neutrons.

With the neutron it was hoped at last to have a projectile which, because it was 'uncharged', would be able to force an entry into the hitherto impenetrable fortress of the atom. In his work 'The Interpretation of the Atom' of 1932, Soddy was one of the first to give his views about the fundamental significance of the neutron as an 'invaluable new weapon' for the transmutation of the atom, possibly for its splitting. One question, of course, remained open: How can this problem be experimentally overcome?

9

National Socialism—Night Falls on German Science

9.1 As Visiting Professor in America

In February 1933 Professor Hahn travelled across the Atlantic once again. Cornell University in Ithaca, New York, had invited the by now world famous radiochemist to give a lecture series. In spite of the uncertain political situation in his home land, Hahn took up this call. On 30 January 1933 Adolf Hitler was made Reich's Chancellor and the National Socialists took over power. *I do not think that his rule will last long*, estimated Hahn of the new political situation, *and I do not have any serious worries about the future*.

So it is that we find Otto Hahn, who acted politically so free of care and unsuspectingly, not completely buried in the questions of the politics of the day during his voyage. Much more was he concentrating on his interests in scientific problems, above all on working out the details of his inaugural lecture 'From the Ponderable to the Imponderable'.

Hahn's introductory lecture, which he delivered in a large auditorium, was granted a complete success. The text was published in the respected American periodical *Science*. Important educational establishments such as Harvard University invited this prominent guest to give further lectures. It was with special delight that Hahn took up the request of Professor Eve to give an invited lecture at Montreal. McGill University acclaimed Hahn with ovations when he spoke about the progress of research into radioactivity in the period between his two visits to Montreal, in 1906 and 1933.

As arranged, Otto Hahn collected together his lectures at Cornell University into a book, *Applied Radiochemistry*. I doubt whether I have read the chapters of any book as carefully or as often as I have in Hahn's *Applied Radiochemistry*, confessed the physicist Glenn T. Seaborg, later Chairman of the United States Atomic Energy Commission. I needed his book like my bible! Hahn's monograph was published in Ithaca, New York, in London, and in Oxford in 1936. A Russian translation was soon published. Oddly enough, this particular publication of Hahn's of a textbook character has never been published in German.

Naturally, the American public wanted to know more about their guest from Germany than just dry scientific pearls of wisdom. The press were interested in how Hahn assessed the political developments in his home land after Hitler's takeover of power. Hahn could not avoid being pushed so directly into the area of politics. From the interviews one could see not only his concern and uncertainty about having to form an opinion about the current politics, but also his lack of concern.

A few years before, on the occasion of a social gathering to which Ernst Feder, publisher of the *Vossiche Zeitung* in Berlin, had been invited, Hahn impressed the guests by his lack of interest in politics. Feder, who as a Jew had to emigrate in 1933, left in his *Diary of a Berlin Commentator*[1] a revealing note: "Professor Hahn, who revealed his political naïvety", came to hear "he who is accustomed to very exact thinking must also apply it to politics". Hahn's indifferent stance on National Socialism would soon have to be soaked in a process of deep rethinking.

On 8 March 1933 Otto Hahn celebrated his fifty fourth birthday in America. In his room he found a large bouquet of flowers, a gesture which the foreigner found kind. Whom were the flowers from? Who, particularly in these parts, could have known about his birthday? The matter was cleared up when two good friends, emigrants from Germany, unexpectedly came to visit him. Professor Rudolf Ladenburg and his wife were the well wishers. From them Otto Hahn learnt depressing news from Germany.

Hitler had hypocritically represented his government as a 'Cabinet of National Unity'. The measure of February 1 turned out to be a transparent manoeuvre to dissolve the Reichstag and to announce 'new elections'. In the eventide of 27 February the Reichstag stood in flames. As it emerged, this case of arson was organised by the Nazis. The new ruler zealously spread the deliberate propaganda that the burning of the Reichstag was the signal torch for a communist uprising. In Germany officials and members of the KPD[2] and SPD[3], and thousands of other opposition groups, were arrested and taken away to concentration camps.

---

[1] *Tagebuch eines Berliner Publizisten.*
[2] *Translator's Note:* The German Communist Party.
[3] *Translator's Note:* The German Socialist Party.

The election for the Reichstag arranged for March 5 became no longer an election according to the civil democratic rules of play. Gangs of thugs terrorised all democratic and anti-fascist progressively minded citizens. In spite of a terror, for propaganda purposes hitherto unparallelled in Germany, only little more than fifty percent of the voters gave the National Socialists their vote. Eighteen percent of the votes went to the Social Democrats and twelve percent to the Communists. Without more ado the seats of the KPD were annulled. Hitler tolerated no more opposition in the Parliament. After the new 'Authorisation Law' the Nazis could now, without regard to the Reichstag, arbitrarily pass laws and enact rulings.

When Rudolf Ladenburg reported these happenings Otto Hahn at first could not find any words. At last the first letter from the home land arrived. How much was the very concerning news to be feared? Lise Meitner had written likewise and lent expression to her worries: "The political situation is very strange, but I greatly hope that it will turn into a sensible road (8 March 1933) .... Here, naturally, one and all stand in the midst of a political revolution. Today is the ceremonial opening of the Reichstag in Potsdam. Last week we received the instruction from the KWG[4] to hoist the black, white, and red swastika flag (21 March 1933).... In the Institute a quite large National Socialist cell has developed..." (3 May 1933).

The notorious law on the restoration of the permanent civil service was enacted by the Nazis on 7 April. Under this hypocritical title was hidden the ruthless purging of the state apparatus of employees who were 'not suitable' or 'non-Aryan', or otherwise offered 'no guarantee of suitability for unqualified entry in the national socialist state'. It was estimated that in the first year of the fascist tyranny 4,000 lawyers, 3,000 doctors, 2,000 permanent civil servants, 2,000 actors and artists, as well as about 2,000 scientists, school teachers, and university lecturers lost their positions of employment on the the ground of the so called Aryan paragraph. If they did not emigrate, prosecution and extermination in concentration camps threatened them.

Lise Meitner, who came from a Jewish family, also had received the usual questionnaire after the enactment of the permanent civil service law. Worried, she asked her fatherly friend Max Planck, who had held the office of President of the Kaiser Wilhelm Society since 1930, and Otto Hahn for advice. Both advised her to wait and see, for no danger should threaten her as a foreigner for the time being.

Bad news, which reached Otto Hahn like the messengers of Job's afflictions, piled up. Professor James Franck, who was friends with Hahn, in solidarity with his Jewish university colleagues laid down his chair at the University of Göttingen and emigrated to the USA. For his departure from Göttingen's railway station, an enormous crowd of people gathered.

---

[4] *Translator's Note:* The Kaiser Wilhelm Gesellschaft.

The farewell which students and the body of lecturers prepared for the departing Franck resembled a demonstration.

The Professors Herbert Freundlich and Michael Polanyi, Fritz Haber's closest coworkers, were dismissed. Albert Einstein, Director of the Kaiser Wilhelm Institute for Physics, had already left Germany in 1932 on the grounds of intolerably increasing anti-semitic attacks, even death threats, and announced his resignation from the Prussian Academy of Science. Fritz Haber, also a Jew, sheltered behind his 'service' in the World War from the otherwise inevitable dismissal. But Haber's belief in Germany was destroyed. He found bitter, but proud, words with which to substantiate his letter of resignation to the Ministry.

"My tradition demands of me in my scientific office that in the choice of coworkers I take into account only the technical and personal qualities of the applicant, without enquiring into the nature of their breeding. You will expect no other way of thinking in a man who has been in charge of his university life for the last thirty nine years, and you will understand that to him the pride with which he has served his German home land all the length of his life now leads him to this request for a move into retirement."

9.2   Hahn Practises Solidarity

In June 1933 Hahn's lectures in Ithaca came to an end. As the thankyou the university organised for its visiting professor a circular tour through the most beautiful regional landscapes of the United States. Hahn hesitated, for he thought it right to return to his institute in Berlin immediately. While he thought about the pros and cons two letters reached him from Planck and Haber, who urgently requested him to to undertake the direction of the orphaned Kaiser Wilhelm Institute for Physical Chemistry and Electrochemistry. Now the decision was made. On 24 June 1933 Hahn travelled back to Germany.

Back in Berlin once again (on 6 July), Otto Hahn took his first measures in his institute and as President of the Kaiser Wilhelm Society. Hahn was startled when he saw Max Planck's face grey from deep worries. The seventy five year old found himself in a difficult situation. The highest maxim of his actions had always been his most profound humanitarian posture. His view of the world stood in implacable opposition to the ideology of national socialism. As the leading representative of the Kaiser Wilhelm Institute he, on the other hand, could not openly appear to be against the new ruler without the loss of his position.

The constant financial difficulties, caused by economic crises and inflation, had brought the Kaiser Wilhelm Society to the edge of dissolving itself. In the yearly report for 1932–33 it says of this, "the Society has the confident hope that the government of national revolution, which fully recognises the significance of pure scientific research for the good of the German fatherland and its position amongst the people of the world, might find the ways and means to move the Society into a situation to support the Institute... ". The help requested was granted, yet not without self-interest. First of all the Reich's Minister of the Interior and Prussian Minister for Science, Art, and Culture, Bernhard Rust, personally made changes to the organs of management of the Society, in particular through the filling of senatorial positions.

Max Planck wanted to remain true to the principles he had practised in the years of the First World War, "to stay the course and work on". To his regret he had to recognise that keeping on working was no longer possible if capable scientists were chased away, hunted, and incarcerated, as had been happening for many months. He told Otto Hahn of his futile efforts to order a stop to this development. To this end he counted on an 'audience' with the Reich's Chancellor; Hitler got carried away with meaningless empty words, finally replying to Planck's repeated objections with a rage, so that nothing else was left than to remain silent and bid farewell.

Otto Hahn followed Max Planck's report with increasing agitation. "We must do something", he said, after careful reflection. "I suggest a joint protest with recognised scientists who have not been affected by this Aryan Paragraph. A campaign against the dismissal and discrimination of our Jewish professional colleagues. I know many who would instantly join such a campaign".

Tired, Planck waved it aside. "If today thirty professors stood against the action of the government, then tomorrow there would be a hundred and fifty who would declare their solidarity with Hitler, because they want to keep their positions".

Taking over in an acting capacity the direction of the Haber Institute to the end of October 1933 brought duties and responsibilities for Hahn. *I made an effort especially to tone down the severity of those in power. I can not bring about any change in the situation.* In a few cases he succeeded in obtaining financial support for dismissed colleagues; in other cases he could help them in the search for a new place of work.

After years of bad heart trouble, Fritz Haber took up an invitation from the University of Cambridge and emigrated to England. But he then felt a stranger. Rutherford, who taught in the same university, declined to shake the hand of The Initiator of the Poison Gas War, as he called him.

Otto Hahn was deeply worried whenever he thought about the uncertain fate of his loyal friend Lise Meitner. She had shown him that discriminatory questionnaire which had to be filled in in accordance with the law on the restoration of the permanent civil service. She was threatened with at least dismissal from the University of Berlin, at which she had held an extraordinary professorship after years of lecturing. With a petition to the Ministry dated 27 August 1933 Otto Hahn sought to avert the threatened dismissal of his colleague. He set out Lise Meitner's professional qualities as follows. 'In her field she enjoys international acclaim and is of the same rank as the French Nobel Prize Winner Marie Curie. *The lectures on atomic physics and radium research which Professor Lise Meitner has delivered at the University for some years could hardly be given by any other dozent.*

However, Hahn's appeal went unheeded. Max Planck was also granted no success when on 30 August he emphasised in his petition to the Ministry that Professor Meitner was 'an authority of the first rank' and together with Otto Hahn represents 'the soul of the Kaiser Wilhelm Institute for Chemistry', the two researchers to whom its world wide high standing was owed. The Ministry thought that no exception could be made. According to the judgement of the appropriate advisors, Lise Meitner was 'one hundred percent non-Aryan' and consequently 'unprotected'.

Lise Meitner's dismissal from the University of Berlin followed on 6 September 1933. In addition to herself and her good friend Peter Pringsheim, a further forty five scientists received a disreputable notice of dismissal. Altogether, the University of Berlin lost more than two hundred of its best teachers. For the majority their termination meant unemployment and loss of their occupation.

It speaks of Hahn's sense of solidarity with his colleagues discriminated against that he no longer wished to belong to the teaching staff of the university under these circumstances. On 31 January 1934 Hahn informed the University of Berlin of his decision to leave the teaching staff at the end of the winter semester. As grounds Hahn referred to his position as Director of the Kaiser Wilhelm Institute for Chemistry, his activity in the Kaiser Wilhelm Society, and his membership of the Prussian Academy of Sciences making too many demands . . . .

In spite of this preventitive measure Otto Hahn had no choice but to fill in the new questionnaire. Also he had to give 'precise information about descent' about himself and his wife and by his signature confirm that "I understand that by knowingly giving false information I may expect instant dismissal, or the contest of my employment by the department of criminal proceedings with the view to dismissal from service".

Even that was not enough. The Prussian Academy of Sciences demanded of him the compulsory "pledge of loyalty, in accordance with the decree of 18 March 1937, of the Reich's Minister for Science, Education, and Culture".

9.3   Commemoration for Fritz Haber

After a year of national socialist rule Fritz Haber died in Basel, completely broken, on 28 January 1934. A life full of hope, scientific success, mistakes, and doubtless also human tragedy, had come to an end. Germany's press, brought into line, made not a single mention of Haber's demise. Max von Laue alone had the courage to write an obituary for Fritz Haber, which he did with embarrassment. *Naturwissenschaften* of 16 February 1934, carried von Laue's text on its front page: "Themistocles did not enter history as an exile from the court of the King of Persia, but as the victor of Salamis". Haber will—so von Laue declared—enter history as the brilliant inventor who by the use of the synthesis of ammonia plucked 'bread out of the air'.

In the Prussian Academy of Sciences Professor Max Bodenstein considered the scientific achievements of Haber, the Academy's member. His speech was published in the Russian journal *Priroda* ('Nature'), which described the death of the German academic as a 'loss for science'.

One year later, on the occasion of the first anniversary of Fritz Haber's death, Max Planck wanted to hold a commemoration. When the appropriate authorities learnt of it Reich's Minister Bernhard Rust forbade all university members from participating. In the Minister's ban of 15 January 1935 the planned celebration was said to be "provocation to the National Socialist State". Planck intervened. Rust deigned to make a compromise, "taking into consideration that the foreign and national press has already become aware of the matter". Only a purely private celebration by the Kaiser Wilhelm Society was approved, about which the press was allowed to publish nothing.

Professor Karl Friedrich Bonhoeffer of Leipzig, one of the speakers named in the programme, received a ban on speaking. If he were to refuse he would have to count on being dismissed. In his distress he wriggled his way round Otto Hahn, who promised him personally to read his lecture.

The very next day the Rector of the University of Berlin informed him that he, too, could not speak. But Otto Hahn, who had expected something similar, had his cunning answer ready. He saw that he was unable to follow this instruction of the Rector of a university to which he had already not belonged for a year.

On the morning 29 January 1935 Max Planck and Otto Hahn went to the Harnacks' house in Berlin–Dahlem, where the commemoration was to take place. Planck was quite depressed, but he stood firm. "I will carry out this celebration, even if I am hauled out by the police", he had said to Lise Meitner the evening before.

It was to be feared that the Nazis would cordon off the entrance to the Harnacks' house. It is true that on the notice board in the vestibule hung a decree by which it was also forbidden for the members of the Kaiser Wilhelm Institute to visit the celebration. The room was filled to capacity. In the front row sat relatives of Fritz Haber. Carl Bosch and a few representatives of the large chemical firms which used Haber's discovery industrially were present. On the back benches, amongst others from Hahn's institute, Lise Meitner and Fritz Strassmann had taken their places. The celebration went off in a dignified manner. After Max Planck had ended his speech of welcome Otto Hahn stepped up to the lectern, gave a commemorative speech, and afterwards read out the text of Friedrich Bonhoeffer's address.

9.4   German Physics—Excels in the World

The commemoration for Fritz Haber was not the only passive form of opposition of the scientists to the fascist barbarism. Max von Laue had given
one of the first examples when, at the physics meeting in September 1933 in
Würzburg, he delivered a spirited address against the increasingly threatened suppression of the freedom of science. Von Laue was helped on that
occasion by a return to history: "Over three hundred years ago Galileo was
sentenced by the Inquisition, and had to retract his 'heresy' of saying that
the Earth went around the Sun". Max von Laue compared the teachings of
Copernicus and Galileo with the relativity theory of Einstein ostracised by
the Nazis. "And yet it moves!", he exclaimed before the gathered physicists.

Max von Laue's parable was particularly directed against the President
of the Physical Technical Reich's Institution, Johannes Stark. Stark and
his comrade in arms Philipp Lenard were hostile to modern theoretical
physics. Both had been awarded the Nobel Prize for their achievements
in experimental physics, but were later obsessed with scientific fantasies.
Stark's axial atomic model and Lenard's ur-æther theory were some of
those. They had both professed national socialism early on. In 1933 they
needed Hitler's power structure in order to push ahead the previously secretly driven campaign against the 'Jewish' theoretical physics.

"Today Einstein has disappeared from Germany", said Stark, filled with
hatred, at an organised rally. "However, unfortunately his German friends
and patrons have the possibility of apparently making further effect in his
spirit. His chief patron Planck still stands at the summit of the Kaiser Wilhelm Society, and his interpreter and friend, Herr von Laue, is still playing
the role of physical expert witness in the Berlin Academy of Sciences... ".
Lenard and Stark, who denigrated quantum and relativity theory as 'jewish bluff', tried to discriminate against colleagues in the field who thought
otherwise. The attempt was the writing of a four volume textbook of 'German physics' by Philipp Lenard, the first volume of which appeared in 1936.
The attitude of mind of the author and his intention found expression in
the preface, where it says, "The German people have now been fed for
thirty years in science with the achievements of a foreign race and people
and its supporters and successors.... No people has ever begun research
in the natural sciences without being based on the breeding ground of the
already existing achievements of Aryans". On the Reich's Party Day of
1936 Lenard received the NSDAP[5] prize for science.

Even Werner Heisenberg, who received the Nobel Prize in 1933 for his
pioneering works in the field of quantum mechanics, quite suddenly saw
himself as a 'white Jew' in the field of fire of that fanatic. True to Göring's
word, 'Who is a jew, I decide', Stark and Lenard arrogated to themselves

---

[5] *Translator's Note:* German National Socialist Workers' Party.

the classification of German physicists into 'jews', like Einstein, 'white jews', like Planck and Heisenberg, and 'aryans', like themselves.

An exchange of letters of one Oberstudienrat Rosskothen has become well known, in which he complained in 1934 to Alfred Rosenberg, the leading race theoretician of 'The Third Reich', because Heisenberg had been again allowed to teach in a German university. One should get him, who champions and develops further the teachings of Einstein, on the right side of the 'charge of betrayal of the people and race'; "the concentration camp is without doubt the right place for Herr Heisenberg".

Rosenberg, the Reich's Leader of the NSDAP, fundamentally shared this opinion, as emerges from this exchange of letters. "Regrettably it is not possible, in view of the attitude of foreign countries, to teach Professor Heisenberg a sharper lesson or to discipline him as fully as would be desirable".

Such considerations were allowed to drop. Nobel Prize winners were at first allowed a certain protection, as is seen in the example of Heisenberg, altered to the extent that the Nazis found cause to criticise the decisions of the 'jewish influenced Nobel Prize Committee'. The bestowal of the Nobel Peace Prize for 1935 on Carl von Ossietski finally gave them the pretext sought to prohibit in general the acceptance of the Nobel Prize by the people of their own land.

Carl von Ossietski, the editor of *Weltbühne*[6], who had already been arrested in 1933, he had his weekly paper banned, and his writings burnt, received the Nobel Peace Prize when he was in a concentration camp. Protests came in from all round the world. With reluctant detestation he was let 'free', and transferred to a sanatorium, where he died from the consequences of his detention.

From this time on the German scientists were subjected to the strictest regimentation. The Nobel Prize winners Richard Kuhn (for Chemistry in 1938), Adolf Butenandt (for Chemistry in 1939), and Gerhard Domagk (for Medicine in 1939) were forbidden not only the receipt of the Prize, but also were forbidden to convey written thanks. When, nevertheless, they did write to Sweden, the Gestapo intercepted the letters and lamentingly held them up in front of the Prize winners. In the end, they were forced to put their signature to a letter in which they were to protest against the award.

Another 'clandestine' Nobel Prize winner was not to be treated much better—Otto Hahn. On 27 November 1944, the Ministry of Culture presented to him an ultimatum, to decline under all circumstances were he to be nominated for the 1944 Nobel Prize for Chemistry. The single exception should be the Dutch man Peter Debye, appointed Director of the newly built Kaiser Wilhelm Institute for Physics. As a foreigner he could not be barred from receipt of the 1936 Nobel Prize for Chemistry.

---

[6] *The World Stage.*

Planck's time in office as President of the Kaiser Wilhelm Society came to an end in 1936. When his re-election for a further six years was considered, the government gave it to be understood that an extension of Planck's period in office should not be thought of. No 'Governor of Jewry' was desired in so exposed a position, as was said in Stark's jargon. This gave wind of a chance and was seen as an opportunity for the office becoming free. So Reich's Minister Rust hesitated and eventually approved the choice of Carl Bosch as the new President, who had considerable influence in heavy industry.

More and more, the typically always 'free' and 'independent' Kaiser Wilhelm Society had to bend to the pressure of the ruler. New statutes were dictated, and individual research institutes received functions in the guise of allotted four year plans which knew only one goal, armament.

10
Dispute Over the 93rd Element

10.1 Experiment in a Goldfish Pond

Increased scientific work was the only alternative for Otto Hahn and many of his colleagues if they were to escape repressive measures. The last spectacular discoveries in atomic research offered a wealth of new, interesting tasks. One unconditionally wanted to test the effect of the new found 'projectiles', deuterons and neutrons. The Americans were setting up their cyclotron, with which they could accelerate the elementary particles to hitherto unknown energies.

Otto Hahn did not follow this general trend. His former favourite occupation, namely, the tracking down of new radioactive decay products, seemed no longer to be granted a future: *There is little hope of finding new elements of the last series of the periodic table. The most important radioactive substances have all been discovered, but the working with them, applied radiochemistry, has only just begun.* To this conviction, which Professor Hahn presented at a gathering celebrating the centenary of Kekulé, he held fast.

The series of radioelements was thus complete. Otto Hahn would never have hit upon the contrary idea, which again was to occur in the searches for unknown elements, if a highly interesting difference of opinion had not broken out amongst the experts—the dispute about an element with atomic number 93. Such an element would not be able to exist on this Earth, since uranium by its classification within the periodic table is acknowledged to be the last of the ninety two chemical elements.

At the beginning of 1934, Irène and Frédéric Joliot-Curie drew attention to an unexpected research result. In the Paris radium laboratory 'a new type of radioactivity' elucidated was demonstrated. So did the title of their paper proclaim in the *Comptes Rendus* of January 15. This time the French researchers succeeded in getting a first prize. When they bombarded an aluminium foil with alpha rays the foil, after only a short time, displayed a noticeable radioactivity, also, unexpectedly, after the source of radiation had long since been removed far away. This had never before been observed. Investigations showed that the aluminium atom had been changed into a radioactive isotope of phosphorus, which further decayed into silicon. For the first time one had succeeded not in influencing radioactive decay but in artificially triggering it. Was this a further milestone along the road to the unleashing of atomic energy?

This induced, or artificial, radioactivity was produced in other light elements such as boron, beryllium, and magnesium. Lise Meitner was the first to provide a plausible theoretical explanation of this new effect, and demonstrated this artificial radioactivity in a Wilson cloud chamber. Rutherford had already earlier bombarded aluminium with his alpha rays but had not discovered this artificial radioactivity.

Everybody at that time wanted to be the first to explain this latest effect, so it was imperative to study the research reports from Cambridge, Paris, and Berlin–Dahlem. Rutherford, Joliot-Curie, Hahn, Meitner, and their colleagues had arrived at a new concept. To them was added another name in 1934, that of Enrico Fermi. This thirty three year old Italian physicist had plunged into radioactive research with true enthusiasm.

Prompted by the discovery of artificial radioactivity and inspired by Soddy's prediction of the excellent properties of the new projectiles, the neutrons, he had begun systematically, at the Institute of Physics in the University of Rome, to bombard one element after another with neutrons. He had hoped to produce artificial radioactivity with these projectiles, and not solely with alpha particles.

With him worked his colleagues, of the same age, Oscar d'Agostino, Eduardo Amaldi, Emilio Segrè, and the brilliant experimentalist Franco Rasetti, who three years before had shown his ability in Hahn's institute by identifying beryllium radiation.

In their experiments Fermi and his friends went forward methodically. They began with the lightest element, hydrogen, and put it to the mercy of a shower of neutrons. With a counter tube which they had constructed themselves they tested whether after taking away the neutron source—a little tube of radium emanation and beryllium powder, the ends of which were sealed by melting—the irradiated element had become radioactive.

Fermi had it in mind to test all the elements of the periodic table up to uranium. But in the Institute of Chemistry of the university everything needed had not been available for a long time. Laura Fermi, the wife of the physicist, graphically portrayed in her reminiscences how Enrico and his friends, with ordinary shopping baskets under their arms and a shopping list in hand, plundered the dust laden shelves of the Institute, the chemical businesses, and apothecaries, until they saw their requirements fulfilled.

If one pursues the actions of the Italian, they form an exact parallel to the methods of working of Otto Hahn and Lise Meitner. Also Fermi and his coworkers ran down the long corridors of their institute like ones possessed in order to measure the preparations straight after their irradiation had been completed. It had to be possible to do this because a fast decaying element with a half life of only a few seconds might be formed. So they had to take to their heels.

Again and again one saw them turn back with disappointed and slow steps to their places of work. For the first eight substances they could generally detect no artificial radioactivity. But with the ninth element, fluorine, the counter began to tick. As it turned out, other chemical elements would also be activated by neutron irradiation. When they emit beta radiation these activated substances mostly turn into atoms of the next highest element. Fermi had found 'Radioactivity Induced by Neutron Bombardment' according to the title of his research report of 10 April published in *Nature* in May 1934.

Interesting results were hoped for with the last element, uranium. It is the heaviest chemical element occurring on Earth. Its atomic nucleus consists of 92 protons and 146 neutrons altogether. The consequent atomic weight has the value 238. If they succeeded, so Fermi thought, in shooting an extra neutron into the nucleus, an isotope of atomic weight 239 would have to be formed, and if, as usually happened, it decayed by beta radiation, then an element with a nuclear charge of 93 would result. This mortal coil had not yet yielded up this substance. The discovery of such an element was a rousing prospect. It represented a step forward into the unknown by human imagination into a hitherto closed area of matter.

The Italians' delight knew no bounds when they had their first success in their experiments. The irradiated uranium was strongly radioactive, and the radioactive products created emitted as beta rays, as hoped. However, as the investigations showed, they were not identical with the neighbouring elements. The proof of this was easy to show. One needed only to add a small amount of a compound of the transmuted elements and to analyse the mixture chemically. It was believed that the resultant radioelement would be an isotope of thorium (atomic number 90), so the addition of a salt of thorium was sufficient. After the usual chemical separation and purification the total activity would have to be contained in the thorium fraction. If not, then it would be evidence that the unknown isotope would not be thorium. Such radiochemical indications had been often carried out by Otto Hahn and his coworkers, and with the greatest accuracy.

Fermi himself found through repetition of his attempts not the slightest evidence that from the uranium there came isotopes of the known neighbouring elements such as protoactinium, thorium, actinium, or radium. As a result, the radioactive species must have belonged to the elements on the other side of uranium, and therefore must be trans-uranic. In particular, Fermi believed there would have to be a radioactive decay product with a half life of thirteen minutes with the associated entitlement of being ordered as the new element 93.

Fermi cautiously gave his new research report, printed in *Nature* of 16 June 1934, the title 'On the Possible Production of Elements with an Atomic Number Greater than 92'. He was unpleasantly embarrassed by his country's press making the announcement of a 'possible manufacture' of an element 93 and boastfully attributing it to the 'fascist victory of the cultural arena', and speaking of the immense contribution Italian science in the year XIII of fascist time.

This triumphal salvo was repugnant to Enrico Fermi and his friends. Without concerning themselves any further they worked doggedly on. By chance they found out that partitions between the neutron source and the irradiated substance exercised a notable influence upon the strength of the induced radioactivity; for example, an ordinary lead plate such as is used for shielding from radiation.

"Might we once try and see whether an extremely light substance such as paraffin might bring about a similar effect", thought Fermi one day.

From a few wax candles a block was quickly made and in the middle of which a hole was pressed for the neutron source. When Fermi measured the activity of a silver test item irradiated in this way the counter apparatus ticked away at breakneck speed. A wonderful new effect; paraffin had increased the artificially aroused radioactivity of the silver by many times!

Paraffin is a hydrocarbon. The neutrons on their way through the paraffin block encountered countless hydrogen atoms, which have the same mass as neutrons. As a result the fast neutrons were braked, compelled to take a zigzag path. Finally they left the paraffin block with a markedly lower speed. A slow neutron, however, can—and this theory of Fermi's was applauded by his colleagues—encounter an atom of silver with much greater probability, just as if a golf ball which had been speeded up by a powerful stroke eventually rolled leisurely into the hole.

"If Enrico Fermi's theory is right then it must work with ordinary water".

"Let us give it a try", Fermi readily rejoined. His friends made the original suggestion of using the goldfish pond behind the Institute for this experiment.

It must have been a strange sight to see the physicists wading in a goldfish pond with rolled up trousers. They submerged their apparatus, so to say the 'fore-runner' of the light water reactor, in the pond, identified the freshly produced activity and obtained almost the same value as before from their exercise with the paraffin block.

Fermi had already thought further. With the help of this method it must be possible in the near future to produce radioactive substances artificially. Perhaps in such a quantity that they might replace the ever more costly natural radioelements? Thus the method gained a commercial aspect which led Fermi and his colleagues, on 26 October 1934, to register a patent for the process of producing artificial radioactivity with slow or thermal neutrons. The idea that with the help of such a method someday one might be able to obtain atomic energy did not come to the Italian. And yet their discovery was a fundamental step in this direction.

10.2   In Honour of Mendeleev

At the beginning of September 1934 a few German scientists took up an invitation from the Soviet Academy of Sciences to a congress which was to take place on the occasion of the hundredth anniversary of the birth of the Russian chemist Mendeleev. Dmitry I. Mendeleev had—as had also his German colleague Lothar Meyer—through the discovery of the periodicity of the elements, their ordering into a periodic system, and through his predictions of the following elements made a historic contribution.

Amongst the guests from Germany were to be found Otto Hahn and Lise Meitner, the chemists Ida and Walter Noddack, and Alfred Stock from Berlin, Wilhelm Biltz from Hannover, Paul Walden from Rostock, Otto Hönigschmid from München, and Gustav Hüttig from the German University of Prague. Also their colleague Friedrich A. Paneth, who had been driven out of Germany and since then had lived in London, had accepted his invitation.

The international congress was to take place in Leningrad from 10–13 September. Otto Hahn was delighted by a reunion with Professor Chlopin, the founder and Director of the Leningrad Radium Institute, whom he had gotten to know during the Bunsen Meeting of 1932 in München.

In Moscow, the first destination, they went sightseeing in the Soviet metropolis. *Red Square with the Kremlin made a deep impression*, wrote Hahn in his recollections. *What we saw in Russia at that time was very informative for us*. From Moscow they went on to Leningrad, which received its guests in the festal robes of an international congress city.

During the congress Lise Meitner spoke about the structure of the atomic nucleus and about its conformity with the laws of the periodic table. Otto Hahn gave a lecture summarising his latest experimental results in the field of applied radiochemistry. Everything is *strenuous, but very interesting*, Otto Hahn wrote home. *The location of Leningrad is exceptionally beautiful*.

Doubtless the Mendeleev Congress also gave a reflection of the notable quality of Soviet science. For many a guest it was a remarkable realisation. The physical chemist Hüttig spoke in the name of his colleagues when he turned to the natural sciences journal *Priroda* in a letter. In it he expressed his acknowledgment of the 'almost unparalleled fostering' which science had found in the new Russia, and was grateful 'for the boundless friendship' which Soviet scientists had shown their German colleagues.

10.3   An Absurd Theory

Of the countless lectures presented during the Mendeleev Congress one claims our attention, because it gave the decisive impulse to Otto Hahn's new direction of work. On 14 September his colleague the chemist Ida Noddack gave a lecture in Leningrad entitled 'On the Present Day Methods of Predicting Chemical Elements'[1]. The latest discovery of the element number 93 occupied a special place in Frau Noddack's remarks. Not only Professor Fermi in Rome, but also a Czechoslovakian engineer called Odolen Koblič had announced the discovery of this element at almost the same time. Koblič called it bohemium. He isolated the new substance from Joachimsthal pitchblende in which several already well known new elements had been found.

Frau Noddack arrived at the view, on the basis of her own analysis, that bohemium was nothing other than a mixture of vanadium and tungsten compounds. Of a new element there was not a trace. At this, Koblič withdrew his claim.

Fermi's reasoning did not convince the critical Ida Noddack. It had been presumed, this lady chemist noticed, that one could conclude the existence of an element 93 only because neighbouring elements of uranium had not been able to be proved as possible consequential products. Certainly, from the known nuclear physical conversions there always arose only isotopes of the same or neighbouring elements. "One can just as well accept", so her conclusion declared, "that by this new type of nuclear shattering with neutrons considerably more 'nuclear reactions' take place than have been observed up to now. It would be conceivable that by the bombardment of heavier nuclei with neutrons these nuclei should decay into several larger fragments, which admittedly are isotopes of known elements, but they are not neighbours of the irradiated elements".

Actually Ida Noddack's observations, which had appeared in print in September 1934 in the journal *Angewandte Chemie*, must have sprung up like an igniting spark amongst the atomic physicists. They had just increased the theoretical possibilities of smashing the atomic nucleus into a new original variant. But the 'people in the subject area' reacted by holding back, although the physicist Francis William Aston had considered such a splitting of atomic nuclei of the heaviest elements like uranium into nuclei of middling mass for possible, even necessary, release of atomic energy.

---

[1] *'Über gegenwärtige Methoden der Vorhersage chemischer Elemente'*.

Obviously theoretical physics in 1932 had not yet reached a level of development at which it could accept this daring proposition of Ida Noddack. Otto Hahn, speaking later about this, explained ill-naturedly that he had not once dared to cite Frau Noddack's seemingly absurd hypothesis, because otherwise he would have had to be anxious for his calling as a scientist ..... .

Looking back one would have to say that atomic researchers must have been wary, on the grounds of their experiences, about the case of the contentious element 93. But there were, from the early history of discovery, enough examples which could reduce the prejudice. So they had let an epoch-making discovery fall asleep which had already been made in 1934.

That a few prejudices proved to be so long lived certainly contributed, after the 'wonderful year' of 1932, to the surprisingly apathetic attitude of the important authorities in physics. Rutherford held himself back when he was asked for his opinion about this much discussed problem. At a meeting of the British Association in Leicester in the autumn of 1933, he said categorically that in his opinion the commercial extraction of atomic energy would be possible neither now nor later. Also in 1937, shortly before his death, this founder of atomic physics and leader of then the greatest nuclear physics research laboratory held no essentially different opinion. "Whoever sees in the transmutation of the atom a source of energy is talking nonsense".

Einstein, whose word carried no little weight, shared the opinion of his English colleague. "Do you believe that it is possible to set free the enormous amounts of energy which your equations demonstrate by bombarding the atom?" Such questions and similar ones hailed down upon Albert Einstein at a press conference in 1934. "I believe that in practice this is not possible", replied Einstein. "To want to split the atom by bombarding it is rather like shooting birds in the dark in a place where there are only a very few birds".

With such overwhelming reservations from the leading representatives of physics, there is little wonder that the slight voice of the chemist Ida Noddack died away unheard.

Rutherford's and Einstein's doubt, however, did not silence the expectations which public attention tied to the wonder science of atomic energy. Despite all the scepticism, the atomic researchers stayed on the track of this problem.

There was no lack of attention paid to the serious warning to exercise 'necessary caution'. Frédéric Joliot-Curie's forceful words, which he delivered during the Nobel Prize Ceremony of 1935, stick in the memory. The French atomic researcher feared that science, once it had learnt how arbitrarily to build up or smash atoms would also be in the situation where it launched a "nuclear transformation of an explosive character". An inexorable nuclear chain reaction could lead to the self-extermination of the mortal world. "We must then quickly look ahead with the deepest con-

cern at the consequence of unleashing such a fundamental catastrophe",
Joliot-Curie gave us to bear in mind. "Astronomers sometimes observe a
star of average brightness increasing greatly. Since this sudden brightening
of a star is perhaps caused by transmutations of an explosive character,
researchers will, without doubt, seek to put this process into practice, but
we hope with the imposition of the necessary precautions".

10.4   Elements 93, 94, 95, 96 . . . *ad Infinitum?*

Hahn's interest in the strange processes involving the element 93 was re-
vived by Ida Noddack's lecture in Leningrad. To the astonishment of his
coworkers, especially Lise Meitner, who wanted to encourage them to copy
Fermi's experiment, Hahn remained reserved at first. "It is hard to believe.
It is asked that I should go back once more to my old days of searching for
new elements. But I have done that before. Now I press ahead with applied
research."

But then it gripped him. When his earlier coworker Aristid von Grosse
maintained that Fermi's trans-uranium was not a new element, but in
actuality an isotope of the element 91—protoactinium—Hahn's ambitions
were kindled. Who was right, Fermi or Grosse?

'On the Artificial Conversion of Uranium with Neutrons'[2] Hahn and
Meitner called their first work on the uranium problem, which was published on 11 January 1935 in *Naturwissenschaften.* In the text it says,
*Amongst the many radioactive substances produced with neutrons uranium
claims a special interest. For Fermi, Rasetti, and d'Agostino came to the
conclusion from their common investigation that a radioactive element of
higher atomic number than 92 was formed by the irradiation of uranium
with neutrons . . . . We now have set about a detailed investigation of this
uranium process . . . .*

It was not protoactinium. Hahn and Meitner succeeded in ascertaining
that without any difficulties. Therefore was Fermi right with his trans-
uranium?

At first both researchers believed that with a few experiments they would
soon be able to decide whether they had been wrong about the secret of
trans-uranium. The experiments became ever more complicated. After irradiation one of the uranium substances, laboriously separated, with a
particular half life suddenly proved, after repeated separation, to be more
complex, being made up of several isotopes. That made a definitive statement more difficult, particularly since not only uranium, but also thorium,
would be converted many times through the neutron bombardment.

---

[2] *'Über die künstliche Umwandlung des Urans durch Neutronen'.*

Out of the first experiments of Hahn and Meitner, an extensive research programme was developed. Up to the end of 1938, after four years' work on the uranium problem, there were fourteen publications which testified to the diligence of the pair. Hahn later called the result of this year *an almost tragic conclusion.*

A new coworker with admirable analytical ability was most happily added to the Hahn–Meitner team, Dr.-Ing. Fritz Strassmann, who in 1929— when twenty seven years old—had started work at the Kaiser Wilhelm Institute, and since 1 January 1935 had been assigned as Hahn's assistant. For the first time his name appeared as coauthor of a publication on the uranium problem on 2 August 1935. Further joint papers followed, and the three names of Hahn, Meitner, and Strassmann became synonymous with precision, originality, and imaginativeness in radiochemistry research.

To the known decay series of natural radioactivity was added the hypothetical transmutation series of neutron irradiated uranium which Hahn, Meitner, and Strassmann had continually to correct. It proved itself to be a task of the highest complexity to bring this new decay series into systematic order and to explain the origin of the elements 93, 94, 95, 96, 97—also named eka-rhenium[3], eka-osmium, eka-iridium, eka-platinum, and eka-gold. For in their publications these researchers left no doubt that they had succeeded in proving the existence of the trans-uranics 93 to 97.

When Otto Hahn delivered his lecture 'Natural and Artificial Radioactive Elements of the Last Series of the Periodic Table'[4] in the Physical Society of his home town Frankfurt am Main, he affirmed that these artificial trans-uranics had been proved and produced in his institute. This was also reported in the Frankfurt *General-Anzeiger* in its edition of 10 December 1935 under the headline 'New Elements—Artificially Produced! From the Work Place of the Atomic Researcher'. "As Professor Hahn has found, there are already at least three various such heavy elements which have been artificially produced. The longest lasting has a half-life of three days. The new elements originate naturally always in only extraordinarily minute amounts. Nobody has yet seen them with their eyes ... ".

Nobody has seen them, and yet they are called elements beyond uranium!?

Already in March 1936 Otto Hahn was able to report a transformation product which Fermi had not found, the new uranium isotope 239. Lise Meitner explained the origin of this 23 minute substance by a 'resonance process', through the addition of a neutron to the uranium nucleus.

For Otto Hahn and his coworker there was not the least doubt that this beta emitter must transform itself into the element 93, eka-rhenium. With their comparatively weak preparations, the Berlin investigators were

---

[3] *Translator's Note:* 'eka' — beyond.

[4] *'Natürliche und künstliche radioaktive Elemente der letzten Reihen des periodischen Systems'.*

regrettably unable to prove this transformation product—it was to have been the first pure trans-uranic. From this they concluded that it was a long lived element. They attached no further significance to their discovery because they were convinced that the representative of the element 93 had already been adequately found and identified. It was a tragic mistake.

In June and July 1937, after their interest in these astonishing transformation products had been awakened, Irène Joliot-Curie and her coworker Paul Savitch began their own experiment. The endeavours of the Parisian researchers at first contributed more confusion, in that they isolated a further radioactive substance of 3.5 hours half life, and their publication of 1 August 1937 explained it as an isotope of thorium.

March 1938.   The unknown transformation product of uranium must be ascribed to another element, so the French researchers corrected themselves, for it is separated from thorium. From its properties it could be an element of the rare earths, probably being an isotope of actinium (element 89) if it is not a new trans-uranic with completely unexpected properties.

In a personal letter to Irène Joliot-Curie, Hahn and Meitner sought to make it clear that thorium, in their opinion, did not in any way come into the question. In Paris this objection was accepted and a new explanation was sought desperately.

May 1938.   Joliot-Curie and Savitch subjected the 3.5 hour substance to a more careful purification. This showed them that their suspicion that it might be a matter of actinium was only just maintainable. Much more did the new radioelement possess the characteristics of lanthanum (element 57).

September 1938.   In their latest publication Joliot-Curie and Savitch admitted that they had made no headway at all in the classification of their 3.5 hour substance.

In Hahn's institute they poked fun at the work practices of the French. They called the 3.5 hour substance, sometimes jokingly, sometimes caustically, 'Curiosum'—it had just to be a derivative of Curie. Yet it must be admitted that those in Berlin–Dahlem were subdued, because for the time being they had been able to contribute nothing to the explanation of this transformation product. On the other hand, Hahn, Meitner, and Strassmann found yet another new substance, a long lived transformation product with a half life of sixty days, which they ordered—on certain proof—as a new trans-uranic, as one can gather from their publication of 12 July 1938. But this work is not worth mentioning for this result. What nobody in the least thought at that point was that it was the last joint publication of Otto Hahn and Lise Meitner after over thirty years of working closely together.

10.5   Lise Meitner in Exile

The political skies over Europe darkened further in the year of 1938. In March Hitler annexed Austria and made it a part of Germany. For Lise Meitner and many of her countrymen this presented a serious situation. As 'German citizens' living within the German Reich they were now subject to the régime of the racial laws, from which up till then as foreigners they were protected. Those belonging to the Institute who sympathised with the Nazis denounced Lise Meitner to the Ministry of Culture, "The Jewess endangers the Institute!"

Hahn was profoundly dispirited by this development. He prepared himself in all haste for what he should do. With the help of an application to the Reich's Interior Minister Wilhelm Frick, endorsed by the President Carl Bosch, a legal emigration was wanted to be arranged. In Sweden, Denmark, Holland, and Switzerland, Lise Meitner had friends who would help her further. Lise Meitner and Otto Hahn received authoritative support for the execution of the plans for flight through Dr. Paul Rosbaud, with whom she was friends. Having acted as scientific editor and subject advisor with Verlag von Julius Springer in Berlin since 1936, Rosbaud was responsible for many publications of the publishing house, and looked after the journal *Die Naturwissenschaften*, amongst others. On the instructions of Springer–Verlag Rosbaud established contact with scientists and technologists inside and outside the country. Just one year later it would become known that he had one other client, the British Secret Service. Rosbaud, a convinced anti-fascist, worked for the Secret Service under the cover name of 'the Gryphon'.

Letters and telegrams were hurriedly exchanged. Dirk Coster, a Dutch physicist, came to the Embassy in Berlin, so that if need be Lise Meitner might be able to cross the Dutch border without a visa.

In the midst of this situation marked with tension and nervousness the answer to Bosch's application arrived. It was rejected, as had been feared. Bosch gave Lise Meitner, who to be on the safe side was not staying at her home but had secluded herself in the Berlin hotel 'Adlon', the text of the refusal over the telephone. Hurriedly Lise Meitner put down in shorthand, "... may I tell you that political reservations exist against the granting of a foreign pass to Professor Madam Meitner. It would be undesirable for noted Jews to travel abroad, there to be a representative of German science, or even to be there in their own name and on the basis of their own experience, and by their employment there to work against Germany... ".

Lise Meitner's flight was a stark necessity under these circumstances. With utmost haste Hahn and Rosbaud helped her with piles of essential things. Otto Hahn gave her a brilliant cut diamond ring, an heirloom of his mother, for the direst of emergencies. In the early hours of July 13 his colleague left in the direction of the Dutch border. Those remaining behind were worried about her fate. The longed for telegram brought the first relief, with the agreed codeword which reported the success of her flight. *I will never forget the 13th July, 1938*, recalled Otto Hahn.

A little time later, a short announcement appeared in an official report of the Kaiser Wilhelm Institute, "Prof. Lise Meitner, who belonged to the Institute as a scientific member from 1914, and who led the physical–radioactive division from its beginning, entered upon retirement on the 1st October, 1938".

In Stockholm the 'retired lady' found an opportunity to work in the institute of the physicist Karl M. Siegbahn. But the further progress of her uranium researches, in which she had been so authoritatively involved, she could only follow from afar with melancholy.

To Otto Hahn she later bitterly lamented that he had sent her away, and at the very moment their joint scientific lives' work was approaching a high point. Lise Meitner clearly supported her case with a part of her shorthand annotated letter of Frick the Interior Minister where it said, "A way must surely be found by the KWG for Madam Professor Meitner, especially in view of her qualifications, to remain in Germany and to work in the interests of the society under private circumstances. This view the SS Reichsführer and the Chief of the German Police in the Reich's Ministry of the Interior, in particular, had supported".

Could one have trusted these vague assurances? Experiences with the Nazis' methods spoke against doing so. Hahn believed, at any rate, that for Lise Meitner there existed immediate danger. It was by reason of the very unfortunate accusation against his lady colleague of so many years that he recalled the fate of other Jewish fellow citizens who were not successful in escaping the clutches of the fascists. In his moments of reflection he thought of scientists like Enrico Fermi, who willingly turned their back on their home. The Italian took the award of the Nobel Prize for Physics in 1938 in order to leave fascist Italy for good with his Jewish wife and to emigrate to the USA. Otto Hahn hoped that Lise Meitner, like many other emigrants abroad, would find a new workplace. It was only with horror that he described what would have happened if his colleague had remained and had had to live through the night of November 9–10. During that night mobs went berserk on the streets, burning synagogues and businesses, and the Jewish fellow citizens were hunted as fair game. 'Kristallnacht' was to be the signal for the ruthless extermination of the Jews.

SCIENCE

VOL. 77   FRIDAY, APRIL 28, 1933   No. 2000

FROM THE PONDERABLE TO THE IMPONDERABLE[1]

By Professor OTTO HAHN

DIRECTOR KAISER WILHELM INSTITUTE FOR CHEMISTRY, BERLIN-DAHLEM, NON-RESIDENT LECTURER
IN CHEMISTRY AT CORNELL UNIVERSITY ON THE GEORGE FISHER BAKER FOUNDATION

CHEMISTRY has for its purpose the study of the composition of our material world. Its first task is to determine the simple basic substances—the chemical elements—out of which all other substances are made, and artificially to produce new kinds of substances from these same elements. After its problem had been thus recognized and defined, thanks chiefly to Robert Boyle, chemistry could be spoken of as a "science," striving, in contrast to the direction of earlier efforts, towards an ideal objective through unprejudiced researches. The prerequisite for these researches was the recognition of the fact that the weight of a chemical compound is equal to the sum of the weights of its constituents. We owe to the French chemist Lavoisier the recognition of the full significance and the ingenious application of this law. We have him to thank for introducing the well-known balance as a reliable guide in chemical work, whereby Lavoisier became the true founder of modern chemistry whose victorious march began in the nineteenth century and has continued at a steadily increasing tempo.

To be sure, Lavoisier's immortal services were poorly rewarded by his contemporaries; in 1794, during the confusion of the French Revolution, the Revolutionary Tribunal sent him to the guillotine.

My topic is "From the Ponderable to the Imponderable" in chemistry, in physics, and I might also add, in biology. How far can we extend the limits of our qualitative and quantitative tests of chemical compounds? Are there methods of investigation that are reliable at and beyond the present limits of our balances? What are the lower limits?

As a science develops, its methods are improved and its aids become more and more refined. The

[1] Introductory public lecture.

FIGURE 10.1. Otto Hahn's invited lecture in the USA, 1933.

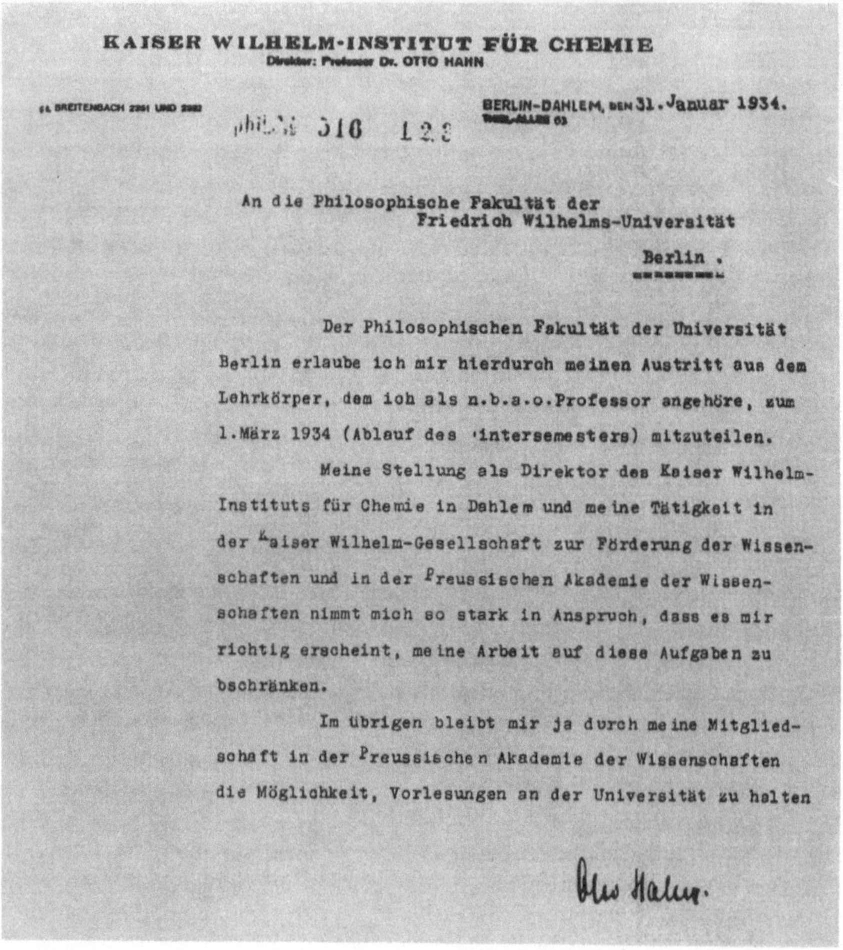

KAISER WILHELM-INSTITUT FÜR CHEMIE
Direktor: Professor Dr. OTTO HAHN

st. BREITENBACH 2281 UND 2282

phil. ?? 316   123

BERLIN-DAHLEM, DEN 31. Januar 1934.
THIEL-ALLEE 63

An die Philosophische Fakultät der
                Friedrich Wilhelms-Universität

                          Berlin .

            Der Philosophischen Fakultät der Universität
Berlin erlaube ich mir hierdurch meinen Austritt aus dem
Lehrkörper, dem ich als n.b.a.o.Professor angehöre, zum
1.März 1934 (Ablauf des Intersemesters) mitzuteilen.

            Meine Stellung als Direktor des Kaiser Wilhelm-
Instituts für Chemie in Dahlem und meine Tätigkeit in
der Kaiser Wilhelm-Gesellschaft zur Förderung der Wissenschaften und in der Preussischen Akademie der Wissenschaften nimmt mich so stark in Anspruch, dass es mir
richtig erscheint, meine Arbeit auf diese Aufgaben zu
bschränken.

            Im übrigen bleibt mir ja durch meine Mitglied-
schaft in der Preussischen Akademie der Wissenschaften
die Möglichkeit, Vorlesungen an der Universität zu halten

                                Otto Hahn.

FIGURE 10.2. Dictated out of a sense of solidarity, Otto Hahn's resignation.

I ask the Philosophy Faculty of the University of Berlin to allow me by this letter to announce my departure from the teaching staff to which I belong as n.b.a.o. Professor, to take place with effect from 1 March, 1934 (the occasion of the inter-semester).

My position as Director of the Kaiser Wilhelm Institute for Chemistry in Dahlem and my activity in the Kaiser Wilhelm Society in the promotion of the sciences and in the Prussian Academy of Sciences make so many demands upon me that it seems right to me to limit my work in these tasks.

I would add that through my membership of the Prussian Academy of Sciences there still remains the possibility for me to give lecture courses at the university.

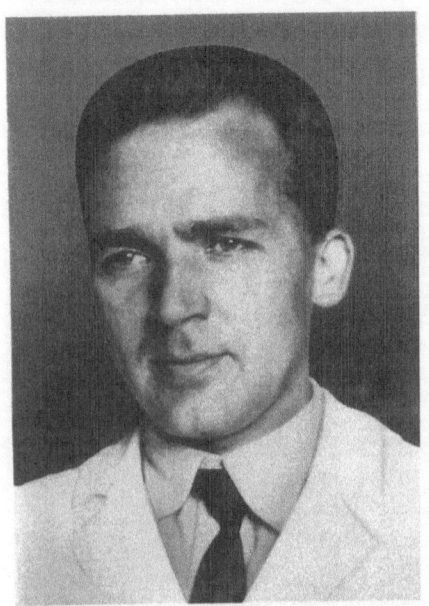

FIGURE 10.3. *Above left:* Fritz Strassmann, 1936. *Above right:* Otto Hahn and Fritz Strassmann explain their laboratory bench with the technical help of which they proved nuclear fission (taken in 1962). *Below:* Otto Hahn's diary; entries in December 1938 (pp. 136–138).

Postverlagsort Leipzig

DIE
NATURWISSENSCHAFTEN

UNTER MITWIRKUNG VON

A. BUTENANDT   P. DEBYE   F. K. DRESCHER-KADEN   H. v. FICKER   R. GRAMMEL
BERLIN-DAHLEM   BERLIN-DAHLEM   GÖTTINGEN   WIEN   STUTTGART

O. HAHN   M. HARTMANN   F. KÖGL   M. v. LAUE   E. v. d. PAHLEN
BERLIN-DAHLEM   BERLIN-DAHLEM   UTRECHT   BERLIN   POTSDAM

F. SAUERBRUCH   H. SPEMANN   H. STILLE   F. v. WETTSTEIN
BERLIN   FREIBURG I. BR.   BERLIN   BERLIN-DAHLEM

HERAUSGEGEBEN VON

FRITZ SÜFFERT

ORGAN DER GESELLSCHAFT DEUTSCHER NATURFORSCHER UND ÄRZTE

UND

ORGAN DER KAISER WILHELM-GESELLSCHAFT ZUR FÖRDERUNG DER WISSENSCHAFTEN

VERLAG VON JULIUS SPRINGER IN BERLIN W 9

HEFT 1 (SEITE 1—16)   6. JANUAR 1939   27. JAHRGANG

Über den Nachweis und das Verhalten der bei der Bestrahlung des Urans mittels Neutronen entstehenden Erdalkalimetalle[1].

Von O. HAHN und F. STRASSMANN, Berlin-Dahlem.

In einer vor kurzem an dieser Stelle erschienenen vorläufigen Mitteilung[2] wurde angegeben, daß bei der Bestrahlung des Urans mittels Neutronen außer den von MEITNER, HAHN und STRASSMANN im einzelnen beschriebenen Trans-Uranen — den Elementen 93 bis 96 — noch eine ganze Anzahl anderer Umwandlungsprodukte entstehen, die ihre Bildung offensichtlich einem sukzessiven zweimaligen α-Strahlenzerfall des vorübergehend entstandenen Urans 239 verdanken. Durch einen solchen Zerfall muß aus dem Element mit der Kernladung 92 ein solches mit der Kernladung 88 entstehen, also ein Radium. In der genannten Mitteilung wurden in einem noch als vorläufig bezeichneten Zerfallsschema 3 derartiger isomerer Radiumisotope mit ungefähr geschätzten Halbwertszeiten und ihren Umwandlungsprodukten, nämlich drei isomeren Actiniumisotopen, angegeben, die ihrerseits offensichtlich in Thorisotope übergehen.

Zugleich wurde auf die zunächst unerwartete Beobachtung hingewiesen, daß diese unter α-Strahlenabspaltung über ein Thorium sich bildenden Radiumisotope nicht nur mit schnellen, sondern auch mit verlangsamten Neutronen entstehen.

Der Schluß, daß es sich bei den Anfangsgliedern dieser drei neuen isomeren Reihen um Radiumisotope handelt, wurde darauf begründet, daß diese Substanzen sich mit Bariumsalzen abscheiden lassen und alle Reaktionen zeigen, die dem Element Barium eigen sind. Alle anderen bekannten Elemente, angefangen von den Trans-Uranen über das Uran, Protactinium, Thorium bis zum Actinium haben andere chemische Eigenschaften als das Barium und lassen sich leicht von ihm trennen. Dasselbe trifft zu für die Elemente unterhalb Radium, also etwa Wismut, Blei, Polonium, Ekacäsium.

Es bleibt also, wenn man das Barium selbst außer Betracht läßt, nur das Radium übrig.

Im folgenden soll kurz die Abscheidung des Isotopengemisches und die Gewinnung der einzelnen

Glieder beschrieben werden. Aus dem Aktivitätsverlauf der einzelnen Isotope ergibt sich ihre Halbwertszeit und lassen sich die daraus entstehenden Folgeprodukte ermitteln. Die letzteren werden in dieser Mitteilung aber im einzelnen noch nicht beschrieben, weil wegen der sehr komplexen Vorgänge — es handelt sich um mindestens 3, wahrscheinlich 4 Reihen mit je 3 Substanzen — die Halbwertszeiten aller Folgeprodukte bisher noch nicht erschöpfend festgestellt werden konnten.

Als Trägersubstanz für die „Radiumisotope" diente naturgemäß immer das Barium. Am nächstliegenden war die Fällung des Bariums als Bariumsulfat, das neben dem Chromat schwerstlösliche Bariumsalz. Nach früheren Erfahrungen und einigen Vorversuchen wurde aber von der Abscheidung der „Radiumisotope" mit Bariumsulfat abgesehen; denn diese Niederschläge reißen neben geringen Mengen Uran nicht unbeträchtliche Mengen von Actinium- und Thoriumisotopen mit, also auch die mutmaßlichen Umwandlungsprodukte der Radiumisotope, und erlauben daher keine Reindarstellung der Ausgangsglieder. Statt der quantitativen, sehr oberflächenreichen Sulfatfällung wurde daher das in starker Salzsäure sehr schwer lösliche Ba-Chlorid als Fällungsmittel gewählt; eine Methode, die sich bestens bewährt hat.

Bei der energetisch nicht leicht zu verstehenden Bildung von Radiumisotopen aus Uran beim Beschießen mit langsamen Neutronen war eine besonders gründliche Bestimmung des chemischen Charakters der neu entstehenden künstlichen Radioelemente unerläßlich. Durch die Abtrennung einzelner analytischer Gruppen von Elementen aus der Lösung des bestrahlten Urans wurde außer der großen Gruppe der Transurane eine Aktivität stets bei den Erdalkalien (Trägersubstanz Ba), den seltenen Erden (Trägersubstanz La) und bei Elementen der vierten Gruppe des Periodischen Systems (Trägersubstanz Zr) gefunden. Eingehender untersucht wurden zunächst die Bariumfällungen, die offensichtlich die Anfangsglieder der beobachteten isomeren Reihen enthielten. Es soll gezeigt werden, daß Transurane, Uran, Protactinium, Thorium und Actinium

[1] Aus dem Kaiser Wilhelm-Institut für Chemie in Berlin-Dahlem. Eingegangen 22. Dezember 1938.
[2] O. HAHN u. F. STRASSMANN, Naturwiss. 26, 756(1938).

FIGURE 10.4. The experts learn about the sensational discovery—uranium is exploded!

[handwritten note] echt amerikanische Übertreibung! 28.5.

THE NEW YORK TIMES,

Atom Explosion Frees 200,000,000 Volts; New Physics Phenomenon Credited to Hahn

By The Associated Press.

WASHINGTON, Jan. 28.—American scientists heard today of a new phenomenon in physics—explosion of atoms with a discharge of 200,000,000 volts of energy.

Theoretical physicists attending a meeting sponsored by the Carnegie new 200,000,000-volt force, which is thirty times more powerful than radium, but pointed to the fact that radium is now the most efficient weapon used for the treatment of cancer. Like radium, it may be twenty or twenty-five years before the phenomenon could be put to

FIGURE 10.5. *Above:* "Pure American exaggeration", Otto Hahn's comment on the announcement in the *New York Times* of 29 January 1939. *Below:* Cutting from the *Deutsche Allgemeine Zeitung*, 15 August 1939.

Die Ausnutzung der Atomenergie

Vom Laboratoriumsversuch zur Uranmaschine — Forschungsergebnisse in Dahlem

Von Dr. habil. S. Flügge, Berlin-Dahlem

Phantastische Energien

Als um die Jahreswende auf Grund der chemischen Ergebnisse von Prof. Hahn die Aufspaltung des Urankerns Gewißheit wurde, stellten wir uns sofort die Frage: Wenn nun bei der Spaltung durch ein auftreffendes Neutron einige Neutronen freigemacht werden, was geschieht dann weiterhin mit diesen Neutronen? Sie haben doch Gelegenheit, andere Urankerne zu spalten; dabei wird wieder jenes Neutron neue Neutronen erzeugen und so fort, solange noch Uran vorhanden ist, das zertrümmert werden kann. Es muß also eine rasch anschwellende Lawine von Neutronen das ganze verfügbare Uran zertrümmern. Es liegt genau das vor, was man in der Chemie eine Kettenreaktion nennt. Damit ist das erreicht, was bisher nie gelungen war: Mit einem einzigen Neutron, das „zündet", wird eine wägbare, ja beliebig große Menge von Uran umgesetzt und dabei Kernenergie freigemacht. Man kann ziemlich genau angeben, wieviel Energie man so gewinnen kann. In der Natur kommt Uran in der Verbindung Uranoxyd vor; sie ist das von Verunreinigungen befreite Erz Uranpechblende, wie es etwa in den Gruben von St. Joachimsthal im Sudetengau gewonnen wird. Ein Kubikmeter dieses Oxyds wiegt 4,2 t und enthält 9000 Billionen Billionen Uranatome. Bei der Spaltung eines Uranatoms werden etwa 3 billionstel Meterkilogramm Energie frei; bei der Umsetzung der ganzen Menge also 27 000 Billionen Meterkilogramm. Da ein Kubikkilometer Wasser eine Billion Kilogramm wiegt, genügt diese Energie, um einen Kubikkilometer Wasser 27 Kilometer hoch zu heben, d. h. also etwa den Wasserinhalt des Wannsees bis in die Stratosphäre emporzuschleudern!

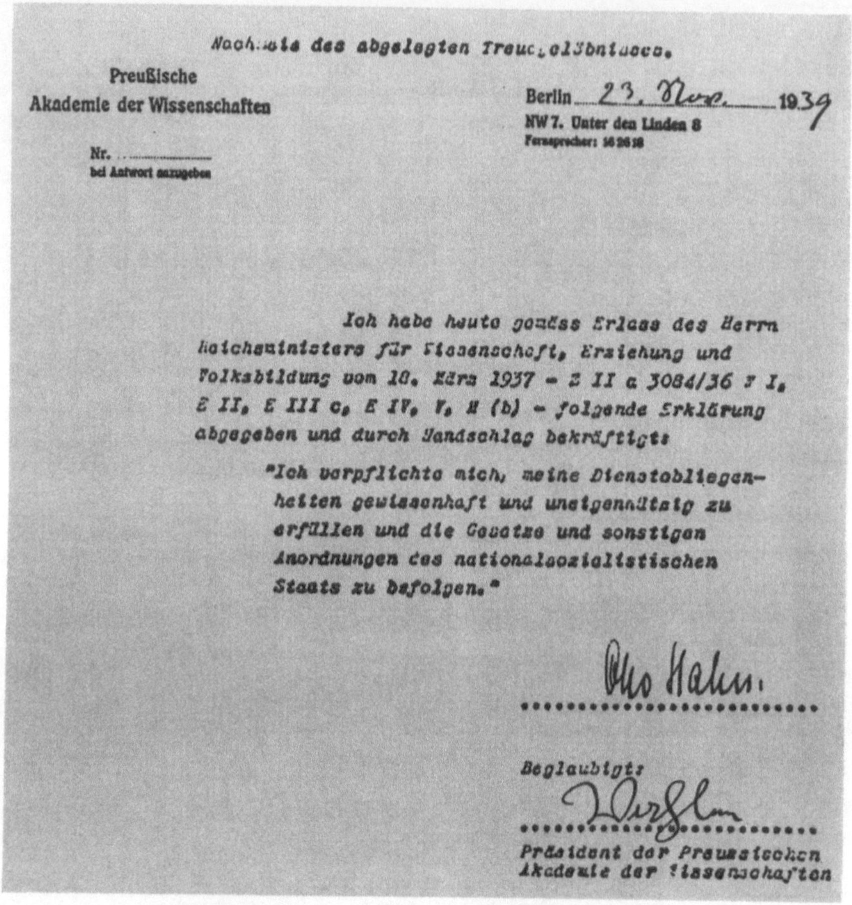

Nachweis des abgelegten Treuegelöbnisses.

Preußische
Akademie der Wissenschaften

Berlin 23. Nov. 1939
NW 7. Unter den Linden 8
Fernsprecher: 16 26 18

Nr. ....
bei Antwort anzugeben

Ich habe heute gemäss Erlass des Herrn Reichsministers für Wissenschaft, Erziehung und Volksbildung vom 18. März 1937 – Z II a 3084/36 Z I, E II, E III c, E IV, V, W (b) – folgende Erklärung abgegeben und durch Handschlag bekräftigt:

"Ich verpflichte mich, meine Dienstobliegenheiten gewissenhaft und uneigennützig zu erfüllen und die Gesetze und sonstigen Anordnungen des nationalsozialistischen Staats zu befolgen."

Otto Hahn.
....

Beglaubigt:
....
Präsident der Preussischen
Akademie der Wissenschaften

FIGURE 10.6. A dubious pledge of loyalty.

Today, in accordance with the decree of the Reich's Minister for Science, Education, and Culture of 18 March 1937, I have delivered and confirmed by a handshake the following declaration:

"I undertake to carry out my duties conscientiously and selflessly and to comply with the laws and various orders of the national socialist state."

FIGURE 10.7. The German V1 super-weapon. The allies feared in the Second World War that Hitler was also able to bring an atom bomb into action.

Secret

Results of Discussion

of the 2nd scientific meeting
of the 'nuclear physics' workgroup
(Reich's Research Office—Military Weapons Office)
in the Deutsches Forschung Building,
Berlin–Steglitz, Grunewaldstr. 35,
on 26.2.1942 at 11 o'clock

1. Nuclear physics as a weapon — Prof. Dr. Schumann
2. Fission of the Uranium nucleus — Prof. Dr. O. Hahn
3. The theoretical foundations for energy production from Uranium fission — Prof. Dr. W. Heisenberg
4. Results of arrangements for energy production so far — Prof. Dr. W. Bothe
5. The necessity of general foundations for research — Prof. Dr. H. Geiger
6. Enrichment of the Uranium isotope — Prof. Dr. K. Klusius
7. The production of heavy water — Prof. Dr. P. Harteck
8. The expansion of the 'nuclear physics' team by the participation of other Reich's departments and industry — Prof. Dr. Esau

Geheim

Vortragsfolge

der 2. wissenschaftlichen Tagung der Arbeitsgemeinschaft
»Kernphysik« (Reichsforschungsrat — Heereswaffenamt)
im Haus der Deutschen Forschung,
Berlin-Steglitz, Grunewaldstr. 35,
am 26. 2. 1942 um 11 Uhr

1. Kernphysik als Waffe — Prof. Dr. Schumann
2. Die Spaltung des Urankernes — Prof. Dr. O. Hahn
3. Die theoretischen Grundlagen für die Energiegewinnung aus der Uranspaltung — Prof. Dr. W. Heisenberg
4. Ergebnisse der bisher untersuchten Anordnungen zur Energiegewinnung — Prof. Dr. W. Bothe
5. Die Notwendigkeit der allgemeinen Grundlagenforschung — Prof. Dr. H. Geiger
6. Anreicherung der Uranisotope — Prof. Dr. K. Clusius
7. Die Gewinnung von Schwerem Wasser — Prof. Dr. P. Harteck
8. Über die Erweiterung der Arbeitsgemeinschaft »Kernphysik« durch Beteiligung anderer Reichsressorts und der Industrie — Prof. Dr. Esau

FIGURE 10.8. *Above:* On 25 April 1945, Otto Hahn was taken prisoner by an American commando unit. *Below:* The country seat of Farm Hall, the place of internment for the German atom researchers in 1945.

11

The Splitting of the Uranium Atom

11.1   The Error Was Tackled with Heaven's Fire

November 1938. In the journal *Die Naturwissenschaften* a further work of Hahn and Strassmann appeared: 'On the Origin of Radium Isotopes from Uranium by Irradiation with Fast and Slowed Neutrons'[1]. The fractionation of the 'curious' 3.5 hour substance carried out again by the two researchers had surprisingly revealed three precipitable radium isotopes with barium salts.

An 'explanation' for the origin of radium from uranium quickly turned up. But it largely sounded muddled, and one had the suspicion that Lise Meitner's orderly hand was missing here. Hahn and Strassmann boldly maintained that the once 'proved' radium (element 88) must have inevitably been produced from uranium (element 92) by four nuclear charges splitting off through the successive emission of two alpha particles. Hahn gave the reasons for that later in the following words: *All chemical elements were excluded except radium, and the carrier barium added in imponderable amounts. Since barium with its lower nuclear charge did not come into the question, there was only radium left.*

Hahn used a lightning visit to Copenhagen on the 13 and 14 of November to let this new theory be sanctioned by Bohr. Bohr, respected by his colleagues for being helpful, was not at all happy about Hahn's request.

---

[1] *'Über die Entstehung von Radiumisotopen aus Uran durch Bestrahlen mit schnellen and verlangsamten Neutronen'.*

Bohr hinted that to him an ejection of two heavy alpha particles from the uranium nucleus by a single neutron was hardly probable, if not to say it appeared to be impossible. Hahn was deaf to it. *I must answer him that there is no other explanation... something other than radium does not come into the question!*

The feverish sensation of being on the track of a new discovery spurred on Hahn's enthusiasm for work. When he arrived in Berlin from his Copenhagen trip at around eight o'clock in the evening, he went straight off to work again. Entries in his notebook diary for the year 1938 reveal this enthusiasm to us, as it says,

> 23 November
> *Physical Society: Von Weizsäcker gave a talk about the universe and the origin of elements. After that in town for a jar, ... after that work again.*
> 24 November
> *Long at work.*
> 25 November
> *A day of great action in work. Attempt at mixed crystals...*
> Sunday, 27 November
> *Edith and Hanno accompanied me at work.*
> 30 November
> *Long trials at work.*

Hahn sent the intermediate results to Lise Meitner by letter, awaiting her congratulations and technical comments. But the opposite happened. *Die Meitnerin* had other worries. "I have become so lazy about letter writing and I also can not believe that anybody enjoys my bad tempered letters", she excused herself on 5 December 1938. "I often feel like a wound up toy doll which does certain things automatically, laughs kindly at them yet has no real life in it". Then further on a cheering sentence, "Your work on uranium is really very interesting... ". And a few lines further on, "I am so sorry that you have your rheumatism again... ". No, in this state he could expect no help from his Lise for the time being.

"We shall prove to Bohr and the others that radium actually does originate from the transformation of uranium", said Otto Hahn to Fritz Strassmann. "Our results must be beyond doubt. Otherwise they will think that Hahn has become old and has been caught fantasising".

With dogged enthusiasm the two chemists set anew to their task. They took complete control of their every move. Hahn and Strassmann were a perfectly matched team. Each experiment was meticulously annotated; Hahn usually wrote down the experimental data and Strassmann the results of the measurements.

The bench on which they carried out their experiments, and on which one of the most significant and at the same time most fateful discoveries was to be made, deserves a closer description. It was an ordinary wooden bench, somewhat over one metre long and three quarters of a metre wide, which had not seen a trace of paint. On it were set out all the utensils needed: a paraffin block about 20 centimetres thick which looked like a round Swiss cheese, in the middle of which was a hole to put the ampoule with the radium–beryllium preparation, the neutron source; near it a siphon bottle with a special vacuum filter Hahn had developed for rapidly separating out the radioactive precipitation products. Little lead boats served as the receptacles of the experiments, the activity of which would be measured precisely in the measuring chamber. A pair of crucible grips lay ready for use. The most space was taken up by the measurement arrangements built according to the ideas of Lise Meitner. Hahn and Meitner used the principle of the Geiger–Müller counter tube, which drove the counter through an amplifier. "This method of counting is at least one hundred times more sensitive for beta rays than the usual electroscope measurements", as the two researchers had already written in 1936 in one of the first publications. Between the pieces of equipment there was a tangle of wires, tubes, coils, and condensers. Under the table was a number of batteries for producing the high voltage for the counter tube. Looked at today, this work bench looks rather like a primitive hobby corner of a radio amateur.

On that winter's day in December 1938 Hahn and Strassmann worked longer than usual. The two wanted exact radioactive proof that the product they found from the irradiated uranium actually was radium. All the reservations of the theoreticians, who Bohr had filled with doubt, were to go to the devil.

Indicator tests, with mesothorium as the radium isotope added to the most varied barium compounds, should resolve to what extent a radium–barium separation was possible under the elegant conditions. Hahn's ten year long experience of dealing with mesothorium paid off. Fritz Strassmann had, in addition, proposed an analytical variant which allowed very pure barium precipitants to be obtained. Instead of precipitating barium (and radium) as sulphates, as usual, which could result in troublesome pollutants, Strassmann successfully tried the method of precipitation from strong hydrochloric acid, which Soddy had already published in 1912. In this way very pure crystalline barium chloride was separated. The choice of this analytical method was to be crucial.

11.2   A Staggering Discovery

Saturday, 17 December
Indicator test Msth 1 + our Ra III, Ra III is not enriched enough, Msth 1 strong!! . . . Exciting Ra–Ba–Msth fractionation.
Sunday, 18 December
Worked for a short time.

Monday, 19 December
... Indicator test: La–Ac (Msth 2 + our AcII). Our Ac different from Msth 2!!

In the late evening of 19 December 1938, shortly before twenty past eleven, the Kaiser Wilhelm Institute on Thielallee was shrouded in the deep dark of the night. Only behind a few windows of the radiochemistry department did a light still burn.

Hahn was sitting down at his desk. There was plenty of time until the next measurement; also Strassmann wanted to take over in half an hour. His glance chanced to fall upon *A Researcher's Prayer* from Sinclair Lewis' novel *Dr. Arrowsmith*. Written on a piece of paper and just about stuck to a piece of cardboard, the text hung over his desk. Hahn skimmed the lines. Never before had its motto had such a legitimacy as at this moment.

> "God grant me clear sight and freedom without haste, God grant me a quiet and relentless hate of all false appearances, of presumptuousness, and of careless and half finished work. God grant me restlessness that I might receive neither sleep nor praise until the results of my enquiries agree with the results of my deliberations, or until I have tackled the discrepancy with the fire of heaven and have vanquished it. God grant me the strength even not to put blind trust in God Himself!"

With a quick grasp Otto Hahn took the laboratory reports into his hand and checked with repeated marks the measurement data taken down by his coworker; the result was unambiguous, and also appeared so wonderful. Deliberately added radium isotopes like mesothorium were in every case separated from the carrier substance barium. But on examination of the neutron irradiated uranium these fractionations did not appear!

Here Lise Meitner had to give further help. With quick resolve Hahn took a sheet of paper and began to write. It would be more a laboratory report than a personal letter (19 December 1938).

There is actually something about the 'radium isotopes' which is so peculiar that we wanted first only to tell you. The half lives of the three isotopes are established well enough; they are separated from all the elements apart from barium..., the fractionation does not work.... The mesothorium was enriched in accordance with the programme, but not the radium. We always come to the following terrible conclusion: our Ra isotope does not behave like Ra, but like Ba. As I said, other elements, transuranium, U, Th, Ac, Pa, Pb, Bi, Po, do not enter into question. I have arranged with Strassmann that we are telling only you first of all. Perhaps you can suggest some fantastic explanation.... Would you think about it, whether there is not some possibility which could be imagined that perhaps there is a Ba isotope with an atomic weight higher than 137?.... I must now get back to the counters.

In the meanwhile Strassmann had come. Hahn gave him the letter to read and his colleague wrote a friendly greeting underneath.

Tuesday, 20 December
... *evening: Christmas celebration in the Institute.*
Wednesday, 21 December
·... *Holidays have begun at the Institute. Strassmann + I write up our on-going work.*

It was eight o'clock in the evening. Owing to the forthcoming Christmas Day the doors of the Institute remained closed for normal business. Strassmann made his rounds and switched off the counter apparatus. The last indicator experiment would have to be postponed until the day after Christmas.

Director Hahn worked on the text of the publication. Initially the title was originally planned to be 'On the Proof and Behaviour of the Radium Isotopes Arising from the Irradiation of Uranium by Neutrons'[2]. After the last experiments, even of those not yet concluded, such a heading no longer seemed maintainable. As is well known, Hahn and Strassmann called it 'Radium Isotopes with the Characteristics of Barium'[3]. Such a curious sounding name hardly gave a right account of the facts of the matter, but should one publicise it? As he honestly admitted, Hahn did not want to draw the critical crossfire of the nuclear physicists again.

Plagued with this kind of doubt, Otto Hahn sought the advice of his assistant. Fritz Strassmann fell upon the answer without difficulty, "It is beyond question that we have found barium, therefore it is barium and not radium!" Hahn, always hesitant, then finally exchanged the concept of radium isotope for alkaline earth. Radium belonged, as did barium, to the group of alkali earths. Doing so always left a door open, in case the nuclear physicists were to drive the two chemists into a corner with their criticisms.

Hahn's careful doubt became clear in a further letter of 21 December to Lise Meitner, expressing it as: *Now we are writing up together our Ra–Ba tests from yesterday....* *After our Ra–Ba tests we must conclude that the three isotopes studied are not Ra, but from the standpoint of chemists are Ba.... We can not hush up our results even if perhaps they are absurd physically.* And finally the sorrowful addition: *How beautiful and exciting it would now be if we had been able to do our work together as before....*

The feeling, the remarkable feeling, if also at first inexplicable, to have made the discovery, probably before other researchers, such as Irène Joliot-Curie, who were standing right on top of it, drove Otto Hahn to an unusual step. Without waiting for the last conclusive indicator tests he decided to

---

[2] *'Über den Nachweis und das Verhalten der bei der Bestrahlung des Urans mittels Neutronen entstehenden Radium-Isotope'.*

[3] *'Radiumisotope mit den Eigenschaften des Bariums'.*

publish quickly the unusual results. He appealed to the editors of *Natur-wissenschaften*, for the next issue he needed an unconditional place for an urgent report.

> Thursday, 22 December
> *Strassmann + I (+ Bohne) finished writing our 'Ba work'.*
> *Dr. Rosbaud collected it in the evening for* Naturwissenschaften.
> Friday, 23 December
> *Still in the laboratory. Work going somewhat better. Lise has written to Kungälv...*

The following days went by with no little excitement. Finally the answer arrived from Lise Meitner, who had received a copy of the manuscript for *Naturwissenschaften*. "Your radium results are very puzzling. A process which runs on slow neutrons and leads to Barium! Incidentally are you quite sure... ?"

As far as the proof of barium was concerned, Hahn and Strassmann trusted their results. But they only ventured to be cautious about their interpretation. In the text of the first publication one finds the characteristic sentence, in which the authors *only hesitantly publish these peculiar results*. No less cautious is the conclusion: *As chemists we really must say the new objects are not a question of radium, but of barium.*

But must one boldly assert that the uranium atom is exploded into one or more elements of intermediate mass such as barium? For the time being Hahn did not risk a storm over the firmly established bastions of nuclear physics, but restricted himself in the meanwhile to the sentence which has become famous:

As chemists we must ... instead of Ra, Ac, Th insert the symbols Ba, La, Ce. As physics is in certain ways on intimate terms with 'nuclear chemists', on the basis of all previous experience of nuclear physics we can not quite make up our minds about such an inconsistent leap. It could perhaps be a series of peculiar accidents that our results have simulated.

Seeking advice, Hahn turned to his former lady colleague and asked in his letter of December 28 for her opinion of his 'theory'. *I would just like to write a few quick things about my Ba pipe dreams.... Could it be possible that the uranium 239 breaks up into a Ba and a Ma? A Ba 138 and a Ma 101 would result in 239. The exact pairing of the atomic numbers does not matter. It could also be 136 + 103 or something similar. The atomic numbers do not turn up naturally in nature. So one must let a few neutrons turn into protons in order that the charges might come out right. Is that energetically possible?*

And so the great practical man Hahn moved onto the treacherous ice of atomic theory, his hypothesis that uranium would split into barium and masurium (technetium) was—false. Theoretically there could be such a splitting if in general one thought only in terms of nuclear charge, that is, the atomic number. Lise Meitner wrote thus on 1 January 1939. The work sent she had "very carefully read and considered whether it could be energetically possible that such a heavy nucleus might explode. Everything indicates to me that your hypothesis that Ba and Ma are produced is impossible on a variety of grounds... ". The criticism of the experienced physicist aroused further doubt in Otto Hahn. On 5 January 1939 he wrote to her that he was *no longer sure today, even apprehensive about the Ba, whether perhaps it is not in fact Ra*.... For that reason Lise Meitner's letter of 3 January must have come to him as something of a relief. "I am now fairly sure that you really do have a break up into Ba, and find that to be a wonderful result on which I heartily congratulate you and Strassmann". That was the confirmation longed for.

One knows in this context that there was no such word for 'nuclear fission' or 'atom smashing', so it is no surprise that in Hahn and Strassmann's by now historic work of 22 December 1938, which was printed in Issue No. 1 of *Naturwissenschaften*, that no such term was to be found. Much more was it the publication of a routine report of an analysis.

Only after the conclusion of the last indicator test, and after Lise Meitner's as it were 'green light' with her communication of 3 January 1939, Otto Hahn and Fritz Strassmann regained their self-confidence. In their next publication of 28 January—which appeared in Issue No. 6 of *Naturwissenschaften* of 10 February 1939—they characterised the observed effect as 'uranium fission', because they were now also sure that their proof was incontestable that the uranium atom explodes and that barium is produced in the process.

Nobody apart from Lise Meitner knew better, because of her long years of experience, how complicated the investigations for a chemical proof of nuclear fission were. Under the elegant test conditions only a few thousand atoms of barium were produced, just a trillionth of a gramme. Also in later years, when Lise Meitner wrote *On and Off the Track of Nuclear Energy*[4], she did not hide her admiration. "I must emphasise", she wrote in 1963, "that this proof with such a low presence of the identifying preparation was in actuality a masterpiece of radiochemistry, in which at that time hardly anybody else other than Hahn and Strassmann would have been able to succeed".

One must recall that the fission of uranium is a nuclear physical process. The phenomenon was, however, discovered by chemists because such a process was not possible in nuclear physics, nor was it worth testing. And yet the physical proof of nuclear fission—by the ionising effects of energy rich fission products—is simpler than the laborious path of protracted chemical separations.

We must add that this occurrence of *explosion* of heavy atomic nuclei, especially of uranium, had first been discussed not by nuclear physicists but

---

[4] *Wege und Irrwege der Kernenergie.*

by a woman chemist, Ida Noddack. For that reason the physicist Walther Gerlach had described, not without some self-mockery, the discovery of the fission of uranium as "a black mark in the history of physics". And Hahn once said with enigmatic humour, *I was afraid that Lischen had prohibited uranium fission to me.*

"...It must certainly be a great joy for you and Strassmann that you have made the whole world of physics excited. That is really wonderful!", wrote Lise Meitner to Hahn on 24 February 1939. Her honest delight at her colleagues' success in their research occasionally gained the upper hand over her feeling of bitter disappointment. As a passionate researcher she could only get over it with difficulty that owing to her flight from Germany she could not at first hand have taken part in this epochal discovery which—felt only at the moment—one makes only once in a lifetime. As a physicist she also feared that her name and her scientific calling might diminish. For she had been crucially involved in the incorrect works on the supposed 'trans-uranium'. All those 'newly discovered' elements the other side of uranium, with atomic numbers 93 to 97, had, however—and this gripped the clever, scientific, and by no means unambitious lady—in reality been the debris of the fission of uranium, and isotopes of known elements. But on no account were they trans-uranium.

"You will well understand", she wrote in her letter of 1 January 1939, to Otto Hahn, "that the question of the correctness of trans-uranium also has a very personal side for me. If the entire work of the latter years were to be wrong, then that will not be put down to just one side. I have been partly responsible and must therefore find some way by which to be involved in a withdrawal. You are certainly—in case trans-uranium disappears— in the very best situation for finding a way out yourself.... Not a good recommendation for me".

Otto Hahn responded with understanding to the disappointment of his colleague. Not in vain had he reported the unexpected results to her in December 1938, for he had hoped this Lise might be able to accept this as compensation. The physicists in his own institute remained, in contrast, completely unawares, for to them Hahn and Strassmann had not said the slightest word.

Because of this professional attitude towards Lise Meitner Hahn got into trouble. When it became known that a jewess outside the country had received information about the splitting of the atomic nucleus at first hand, the voices increased who would push him into the political twilight and accuse him of 'betrayal of the race'.

11.3   200 Million Electron Volts

When Lise Meitner received Hahn's post of the 19 and 20 December she was to be found in the southern Swedish health resort of Kungälv, not far from Göteborg. She stayed there for Christmas, her first time far from home in exile. Her nephew, the physicist Otto R. Frisch, himself a refugee, and who had found a position with Niels Bohr in Copenhagen, visited her. Both quickly got down to talking shop after their first greeting.

Animatedly Frisch described a large new magnet which he wanted to build, so that Lise Meitner had some trouble in steering his attention round to Otto Hahn's letter. After the first of it, he immediately appeared to be fascinated by Hahn and Strassmann's discovery. Yes, it might be possible. Like a drop of liquid filled to bursting, the uranium nucleus would have to divide itself at the slightest stimulus. Bohr had recently spoken about a 'liquid drop model' when he was looking for a clear comparison for the condition of a heavy nucleus. Also a drop of water divides itself into two smaller drops of approximately the same size at a certain critical value of its surface tension, and does not split into innumerable small drops. It must happen in the same way that a uranium atom bombarded by a neutron becomes unstable and splits in two. Biologists call this process in a living cell 'fission', so Frisch remembered. This concept became established as part of the language for the splitting of the uranium atom.

If one of the fragments—as had been found by Hahn and Strassmann—was barium with a nuclear charge of 56, then the other must be none other than the rare gas krypton, of nuclear charge 36, and it could not be masurium as Hahn had thought. The two together added up to 92—the nuclear charge of uranium. The correctness of this assumption could soon be confirmed experimentally. Hahn and Strassmann did indeed also find, as a result of Meitner and Frisch's tip, an isotope of krypton amongst the fission products of the uranium.

From the loss of mass, which occurred in just such a fission, Meitner and Frisch immediately calculated a sum of energy which would inevitably have to be released, as equivalent to that from Einstein's equation $E = mc^2$. They checked their rough calculation again and again because the sum found seemed unbelievable, 200 million electron volts! Such an energy from nuclear transformation processes hitherto had not been observed. For a better comparison, 1 atom of coal releases in its burning an energy of about 2 electron volts (eV); 1 atom of uranium in its fission therefore releases one hundred million times as much! At any rate this energy would be enough, so Meitner and Frisch calculated, to make a small grain of sand visibly hop up. These enormous numbers first come to life if we put them in terms of things which we can comprehend in our world. As a result, the release of energy by the complete fission of 1 gramme of uranium is around 5×10^{23} million electron volts (MeV), or, in other terms, some 20 million kilocalories. In order to generate this amount of heat one would have to burn 2.5 tons of high grade coal. In still simpler terms, with the fission of uranium an amount of energy will be released which is two and a half million times larger than the burning of coal.

At the beginning of January 1939 Lise Meitner travelled back to Stockholm. Frisch resumed his work again in Copenhagen. He reported to Bohr, who was making preparations for a journey overseas, this sensational piece of news on 3 January. For a while Bohr seemed to be speechless. Then he struck himself on the head. "How could we have overlooked that for so long? What kind of idiots have we all been?!" Before they had parted, Meitner and Frisch agreed to publish their ideas about the exploding uranium nucleus in a short paper entitled 'The Decomposition of Uranium by Neutrons. A New Kind of Nuclear Reaction'.

This manuscript—the first interpretation of the phenomenon of nuclear fission, today just as classic an article as that of Hahn and Strassmann—Meitner and Frisch worked on in unusual circumstances. The English text was laboriously agreed upon by telephone between Stockholm and Copenhagen, and on 16 January was submitted to the English journal *Nature*, which carried this contribution in their issue of 11 February 1939. The editorial department had not been asked by the authors for an especially quick publication. Doubtless, had Meitner and Frisch requested that, they would have foreseen the events which followed, rushing upon each other.

11.4   "Real American Exaggeration"

Lost in thought, by the skin of his teeth Niels Bohr had caught the passenger steamer Drottningholm, which was to take him to the United States. The physicist was taking up an invitation to visiting lectures at Princeton University. He was accompanied by his third son, Erik, and his visiting assistant, Leon Rosenfeld, a theoretical physicist from Belgium. During the crossing Bohr and Rosenfeld busied themselves almost without interruption on the 'Hahn effect', that unexpected sundering apart of the heavy uranium atom. These considerations would be displaced only by the worry about the political developments and peace in Europe. To the passengers who observed the Danish physicist on board, he gave the impression that he carried a heavy burden around with him.

Friends and former pupils had gathered in New York at the pier of the Swedish shipping line in order to welcome Niels Bohr on January 16. Amongst them he also found Enrico Fermi and his wife. Bohr had greeted the Fermis in his hospitable home on their journey from Sweden to America. Fermi now worked at Columbia University in New York.

While the customs formalities were being conducted Bohr took his former pupil John Archibald Wheeler, Professor at the university in Princeton, to one side in order to tell him about the world shaking piece of news, uranium fission, whispering in hurried sentences. Wheeler remembered later that the few words had given him a real shock, for he grasped in a flash the menacing consequences. Rosenfeld also could not keep his knowledge to himself.

On 26 January the 5th Washington Conference on Theoretical Physics was to take place. Bohr, having been invited to it, wanted to use the opportunity to make a few remarks about the results of the work of Hahn and Strassmann and about Meitner and Frisch's theory. Shortly before his departure from Copenhagen he had agreed with Frisch not to say anything about the fission of the uranium atom earlier than when, respectively, the work of Hahn and Strassmann appeared and Frisch had arranged the physical dispatch of his contribution. Bohr now remembered these undertakings and reproached himself for not having kept his mouth shut on his arrival.

What he had feared now took place. Before the beginning of the Washington Conference the sensational discovery was already being recounted by a dozen physicists behind a held up hand. Bohr, unhappy about it, telephoned Copenhagen to encourage Frisch to hurry up. The thought that the proper priority of discovery could not be maintained because of his rashness filled Bohr with concern.

In the meanwhile, before the beginning of the meeting, Issue No. 1 of *Naturwissenschaften* for 1939 circulated amongst the participants. Bohr could now speak openly and enquired whether he might make an announcement of the greatest importance. The monotonous course of the Conference up to then took a dramatic turn. Many physicists jumped out of their seats as if given an electric shock before Bohr had gotten to the end of speaking. In their business suits they stormed into their laboratories to follow through the 'oversleeping' discovery of the century. Whoever had travelled from afar glued himself to the nearest telephone to give instructions for the appropriate experiments.

Bohr and Fermi were invited to take part in one of these experiments in the Carnegie Institution. In the midnight hours the physicists, entranced, watched the oscilloscope whose light green pulse indicated the fission energy. Fermi was asked why he had not already noticed the fission of uranium in 1934. The energy rich 100 million electron volt fission fragments of uranium must also have been detectable by his primitive counter. The Italian smacked himself on his head. Of course! But at that time he had put a foil between the irradiated uranium and the counter in order to exclude the natural radioactivity of the uranium. However, even a gossamer thin foil would be enough to absorb the fission debris. So nuclear fission had had to remain a secret of nature in those days.

On the next day, 27 January 1939, the meeting's participants heard that not only in Washington's Carnegie Institute, but also at the universities in Baltimore, Berkeley, and in Columbia University, New York, nuclear fission had been confirmed. Thanks to Bohr and Rosenfeld's indiscretion the experiment was carried out at Columbia University as early as 25 January. Independently of these works of the American physicists, Joliot-Curie in Paris and Frisch in Copenhagen at about the same time were successful in their experiments for the physical proof of fission.

As the news service *Science Service* first reported the Americans' successful experiments on January 25 and 26, "The explosion of the uranium atom is tremendously powerful, even if the 200 million electron volts is not sufficient to provide a household light with power. But even now there are "prophesies that one will soon be able to power great ocean steamers with the energy of the atom... or to turn atomic energy into a kind of super-explosive in weapons" ".

Then the daily press brought the first report, "Atom Explosion Frees 200 000 000 Volts; New Physics Phenomenon Credited to Hahn". The *New York Times* of Sunday, 29 January 1939, blazed with this headline. When Otto Hahn was later sent this announcement of the production of 'millions of volts' (to be correct, electron volts), he added to it the note *real American exaggeration.*

Many press reports sound like the announcement of a catastrophe. Such was there in *Time Magazine* of 13 March 1939, "Six weeks ago a report reached the United States about an atomic explosion which took place in a Berlin laboratory—the most violent ever produced by man". Today one thinks of the ominous mushroom cloud of an atom bomb, many kilometres high. What the readers imagined is not known. At any rate, however, they had to remember an announcement of 6 February. It said that it had been only a matter of "an explosion at the microscopic level, without any flash, and without any bang... , in actuality not even enough to blow a fly off the wall".

11.5   Indispensable Chain Reaction

Numerous scientists came to know in January 1939 that a single uranium atom would be split when bombarded with neutrons. But it was not a self-producing avalanche reaction which could set in motion a wave of nuclear decays as Rutherford and others had feared. It is correct that the match for setting light to the nuclear fire had been found, but the temporary 'fire' went out again as soon as the neutron source was taken away. For the maintenance of the uranium fission, a constantly self-renewing reaction was needed which ran spontaneously without any further supply of energy from outside. The bright stars of the firmament and our Sun are examples that such chain reactions are possible, even are necessary, for the constant release of atomic energy.

This energy supply process running in the universe had already brought the students Fritz Houtermans and his English fellow Robert d'E. Atkinson in 1927 to the idea of explaining the origin of the Sun's energy as nuclear chain reactions, although this energy arises not through the fission of atoms but through the fusion of the lightest elements, hydrogen to helium. A few years later Carl Friedrich von Weizsäcker and Hans A. Bethe interpreted these reactions as a circular process which incessantly played through a chain of reactions in our natural energy delivery man, the Sun.

Houtermans liked to recall that Göttingen student year, and then fondly described the following episode. "I went... for walks with a pretty girl, and when it had become dark the stars then came out one after the other so beautifully. 'How beautifully they shine', my companion called out. I puffed out my chest a little and said, 'Since yesterday I know why they shine...'".

But how must the process of uranium fission be run so that all, and not just a few, of the uranium atoms explode? By thinking about it the investigators came upon the following idea, there must, then, be possible a chain reaction, if further neutrons were to be set free in the fission process, which in their turn could split new atoms of uranium.

In 1932 Houtermans had already voiced a similar thought in his inaugural lecture. A year later the Hungarian physicist Leo Szilard, who had emigrated because of the Nazis, filed a secret patent in which he described theoretically a nuclear chain reaction with the help of neutrons. Szilard believed, though, 'that it must take on an inevitably explosive character', and held back his theory of it. Only to his former champion Rutherford did he give an explanation, remembered Szilard. But the former, so Szilard later openly confessed, chucked it away, and said to him face to face, "That would be sheer insanity. A practical exploitation of atomic energy will never be given".

Might further neutrons be set free in uranium fission? Might one arrive at a chain reaction with their help? Would such a reaction be stoppable, controllable, or devastatingly explosive? Numerous physicists around the world asked themselves this momentous question in the spring of 1939.

In connection with the uranium fission they had discovered, Otto Hahn and Fritz Strassmann first expressed the thought that purely arithmetically a number of neutrons would be able to be emitted. This can be looked up in their second work on nuclear fission, dated 28 January 1939. In it Hahn seemed to feel somewhat disappointed: *We also carried out a search with completely inadequate means, and for that reason without success, to test for such additional neutrons ....*

Frédéric Joliot-Curie and his colleagues Hans von Halban and Lev Kowarski of the Paris 'Laboratory for Nuclear Chemistry' were the first to bring the experimental proof that in uranium fission additional neutrons arise. Their joint work of 8 March was published in *Nature* ten days later. In all fairness it must be said that they would only have been allowed the extremely meagre 'start' of a few hours in this scientific race. For Leo Szilard and his coworker Walter H. Zinn on 3 March likewise carried out this crucial experiment in their laboratory at Columbia University and did so successfully, but they temporarily held back their experimental results. Let us follow the dramatic course of the investigation which Szilard described in the eloquent words, "Everything was by and large arranged so that we had now only to press the button and to observe what turned up on the fluorescent screen. If specks of light appeared there then it meant that neutrons had been emitted in the fission of uranium. But that would indicate that the freeing of atomic energy in our lifetime might be possible ... ".

When they pressed the button they both saw clear specks of light and observed them for about twenty minutes as if captivated. "This very night it was clear to us that the world had set out on a road full of cares", Szilard's commentary pronounced.

Then other researchers also succeeded in this proof, giving rise to a certain victorious confidence that in the fission two to three neutrons were emitted. It seemed only to be a question of the experimental skill to trigger a chain reaction by the fission of uranium. This optimism was to be markedly dampened by a statement of Bohr—published on 15 February 1939 in the *Physical Review*—which put forward the (if also at first uncertain) hypothesis that only the atomic nucleus of the uranium isotope with the atomic number 235 would be fissionable. This statement had a sobering effect because the uranium occurring in nature existed almost entirely of the non-fissionable uranium 238 and only 0.7 percent of uranium 235.

If Bohr was right it would be absolutely necessary to enrich this uranium 235, even to separate it purely by the kilogramme. To even the most experienced experimental physicist this seemed scarcely feasible at that time.

Difficulties had already appeared in the next experiments. At least a whole year would be needed for testing Bohr's hypothesis experimentally— and to be able to confirm it. With a mass spectrograph specially adapted for the purpose, after laborious operations the American Alfred O. Nier separated one two thousandth of a milligram of uranium 235 and 238, and found that indeed only uranium 235 split when bombarded with neutrons.

11.6   60th Birthday Celebratory Volume

There are many examples of natural scientists who are so in the truest sense of the words, such as Planck and Einstein, who distinguish themselves at the start of their career by a brilliant discovery. Usually they live off it for all their lives and erect their scientific work on their discovery. In contrast, Otto Hahn succeeded in his most important scientific achievement in his sixtieth year.

A yardstick for the particular interest in a new discovery is its echo, amongst the experts, in the number of citations in papers. What was the public acknowledgment of uranium fission?

"Their discovery has caused a huge sensation in the whole scientific world", announced Professor Ladenburg on 22 February 1939 in a letter from the United States to his colleague Hahn. "And every laboratory which has the necessary means is now working on the consequences of your discovery.... The issue of the *Physical Review* appearing in the coming week you can regard as a 'Festschrift' for your sixtieth birthday, even if it is not designated so".

8 March 1939. Otto Hahn was sixty years old. The early morning of his special day found this celebrant in an unusual hurry. Outside the front door Max von Laue tooted impatiently on the horn, wanting to pick up the Hahn family in his motor vehicle for the Institute's celebration. There his colleagues and guests were already waiting, amongst whom were Otto von Baeyer, Max Bodenstein, and Gustav Hertz—forty six well wishers altogether. The telephone rang incessantly, the host having to accept congratulation after congratulation. Hahn sought to cut short the conversations with great kindness.

Over 200 congratulatory telegrams, and numerous letters and cards from all over the world arrived during the course of the day. His seventeen year old son, Hanno, who had written out for his father 'The Researcher's Prayer' in calligraphic lettering and put it in a new frame, took the trouble to get everybody to sign it.

In the press the birthday boy was acclaimed as 'the pioneer of radiation therapy' and as 'the discoverer of German radium'. In the list of his scientific achievements the newest discovery of uranium fission was not forgotten. "This revolutionising discovery is now occupying the minds of scientists all over the world", was to be read in a German newspaper.

The specialist journals were in no way inferior to the daily papers. *Current Science*, in the fifth issue of 1939, printed an appreciation from the pen of Lise Meitner. "Otto Hahn's humane and scientific personality is an indivisible whole. A very lively intellectual intuition, a very sound ability, an exceptional and critical ability for observation, an unshakeable dependability and doggedness next to great inner modesty and natural kindness mark the man as they do his work...".

Max von Laue wrote an article honouring his friend, printed on the first page of Issue 10 of *Naturwissenschaften* of 10 March 1939. In the celebratory issue of *Zeitschrift für physikalische Chemie* of March 1939 are all the collected contributions offered for the celebration. Amongst them was to be found an article by two of Hahn's institute's fellow physicists, Siegfried Flügge and Gottfried von Droste, entitled 'Energetic Investigations of the Origins of Barium from Uranium by Neutron Irradiation'[5]. This treatment had already been submitted to the journal on 22 January but had been held back for the celebratory issue.

This circumstance is noteworthy because the authors had reached at practically the same time as Meitner and Frisch an entirely similar interpretation of the nuclear fission, and also had correctly found the fission energy

---

[5] *'Energetische Betrachtungen zu der Entstehung von Barium bei der Neutronenbestrahlung von Uran'.*

as 200 million electron volts. As is well known, the physicists at the same institute as Hahn and Strassmann had not been immediately informed about the surprising experimental results. So Flügge and von Droste first found out about the fission of uranium in the Issue No. 1 of *Naturwissenschaften* of 6 January 1939. First, on Monday 9 January, Hahn's own physicists were informed, as is to be gathered from his notebook. *Discussion about Ba from U with Weizsäcker, Flügge,* etc.. Their sense of resignation lingers on in a report 'On the Discovery of Uranium Fission'[6], which Flügge drew up in 1949, "When at the beginning of January of 1939 we returned from the Christmas holidays everything was already resolved . . . ".

Not all publications on uranium fission deserved to be included in a 'Festschrift for Otto Hahn'. A few awakened contradictory feelings in the participants Hahn and Strassmann. It began with announcements from the USA that American physicists had carried out nuclear fission, and therefore had also 'discovered' it, or at least had been the first to be able to interpret it correctly. Hahn and Strassmann, and Meitner and Frisch fell behind. Bohr, who had foreseen this development, and who from his sense of guilt could not freely speak, sought every opportunity of correcting this impression. But he could not prevent that prejudgement from taking root.

Einstein himself was of the impression that nuclear fission had only been an accidental discovery of Hahn, "who had wrongly interpreted what he had discovered". With justification the scolded one denied it: *the dreadful amount of work and number of chemical investigations which led to the fission nobody can describe as an accidental discovery.* It also was not correct that Hahn and Strassmann had 'wrongly' interpreted the unexpected results. Much more was it that they had taken so long with their careful formulation until they had irrefutable proof in their hands.

If one leafs through the letters which were exchanged between Hahn and Lise Meitner during this time, one is taken by surprise in places by the petty jealousies and vehemence with which Otto Hahn championed his claim to discovery.

Pretty annoyed, Hahn reported in a long letter of 3 March 1939 to *Dear Lise* that now he must gradually put right those who thought that Hahn and Strassmann's priority in uranium fission was threatening slowly to slip out of their hands. The article by Meitner and Frisch in *Nature* the title of which, 'A New Type of Nuclear Reaction', had also contributed to it, for nobody any longer spoke of Hahn and Strassmann, but only about the *new nuclear reaction of Meitner and Frisch.* To judge from an article in *Comptes Rendus*—about which Hahn was more indignant—even Joliot-Curie and her coworker were the real discoverers of nuclear fission.

Such things had capital made out of them during the institute coffee time. A few physicists did not hide their malicious pleasure. It served the

---

[6] *'Zur Entdeckung der Uranspaltung'.*

discovers right because they had not first come to them. Therefore there was also distrust in that Institute. *The most fantastic of our people have turned these things into something political*, Hahn wrote indignantly. Strassmann had been asked in confidence whether the manuscript of the first publication had also been something which Meitner had sent. But the one questioned showed no weakness and strongly repudiated the suggestion . . . .

That we wrote so cautiously in our conscientiousness was, perhaps, wrong, confessed Hahn in his letter. *However, we alone must make our own claim against each physical authority. . . . Indeed, really the only thing which affects me so deeply in this whole ridiculous episode is that never have we worked more conscientiously nor in a more carefully considered way than in these investigations. . . . Had we done a botched up job like Irène Curie then we would have published Barium in November. . . . But now it has all ended with this Hornberger shot. Strassmann and I know that we found the disintegration of uranium without any dependence upon any report or hypothesis.*

Without a doubt these words speak from passion and depth of feeling, from passion and a markedly deep sense of right, if not also very much of making allowances and of intelligent reflection. In Hahn's letters at that time there is mirrored something of the nervousness and distraction which had seized him and many another. Only one seemed not to have been touched by it all, as we can gather from Hahn's letter of 3 March 1939: *The only person who has assessed this matter correctly is Strassmann himself, who—one must say—has come off the worst in every respect.*

11.7   The False Trans-uranics

What complication can arise if in publications the work of predecessors is not correctly cited and whose credit is not recognised or whose name is even suppressed. Otto Hahn had learnt by experience. He was little pleased when he himself was accused of this sin of omission.

Ida Noddack, who in 1934 had described uranium fission as possible in a theoretical speculation, accused Hahn and Strassman in a communication to *Naturwissenschaften* of 10 March 1939. "My criticism of the Fermi investigations have they cited neither in their first nor in their many later publications about the artificial transformation of uranium with neutrons. When this omission was brought to his notice verbally, Otto Hahn declined a citation of my work, evidently because he held my supposition that uranium could perhaps be split into large fragments to be absurd, because such a nuclear reaction seemed at that time to the theoreticians to be impossible . . . Also now, after the disintegration of uranium by neutrons into large fragments has been proved, Otto Hahn refrains from citing my supposition voiced in 1934 that such a process could take place".

With relish Ida Noddack now enumerated the transgressions of the pair, cited from the last work of Hahn and Strassmann of November 1938 in which the authors had reported the discovery of no less than seven trans-uranics and three isotopes of uranium, radium, and actinium. And all had been quite wrong!

Not all 'festschrifts' for Hahn's sixtieth birthday, therefore, brought pure joy. Hahn read the contribution which Rosbaud had sent to him for his opinion, and appeared to be annoyed. *The letter to 'Naturwissenschaften' is unusually unfriendly,* he told his woes to Lise Meitner in a letter. However, Hahn's reply had no charm about it, much more was it fairly ungentlemanly; Strassmann and he had *neither time nor the inclination* to answer such an objection. They believed, so that they might do without it beforehand, that the possibility of the disintegration of heavy atomic nuclei into smaller fragments had been discussed earlier by many others, without any experimental consequences having been drawn. Consequently, they left over their opinion about the entitlement to the claim of Frau Ida Noddack and the nature of her portrayal to her colleagues in the field.

Petty jealousies amongst scientists. Scrapping about priorities. Was it really only this—that as far as the question about the existence of the trans-uranics was concerned Hahn and Strassmann would have to take back their earlier statements step by step? That began in June 1939 with the 'conclusive deletion' of eka-platinum, which turned out to be an isotope of iodine, and of eka-iridium, actually a mixture of tellurium and molybdenum isotopes. But the researchers followed through these corrections themselves and not as a result of outside advice. As for the action of Hahn and Strassmann, they had always published their results and submitted them to the scrutiny of other specialists, and, quite correctly, also in the case of the retraction of their assertions.

The ornate towered building of the trans-uranics had been brought into being in a very short time. The once so decisively determined elements 93 to 97 were nothing other than the fragments of nuclear disintegrations, elements of intermediate atomic weight. Finally, of all the transformation products, only the uranium isotope 239 found by Otto Hahn in March 1936 remained. As a beta radiator it actually decayed into the element 93, eka-rhenium, as was able to be announced in June 1940 by the Americans Edwin McMillan and Philip H. Abelson, honoured with the Nobel Prize. They named the first true transuranic neptunium. Disappointed, Hahn later said, *there went a Nobel Prize.*

In his later lectures, 'The False Transuranics, the History of a Scientific Error'[7], Otto Hahn set out the history of the discovery of the elements 93 to 97 as a very instructive textbook example of scientific mistakes. He did

---

[7] *'Die falschen Transurane, zur Geschichte eines wissenschaftlichen Irrtums'.*

not spare self-criticism in it, which was something omitted by Fermi, who for his false 'transuranium' even received the 1938 Nobel Prize.

The disentangling of the fission products from the disintegration of uranium had, in the meantime, been only one direction which atomic researchers were following in 1939, and not just in the field of work of Hahn and Strassmann. Many groups of researchers in the USA, in the USSR, in France, and in Austria busied themselves with it. Contributions to the clarification of the chemical nature of the fission products were provided by Professor Chlopin and his colleagues at the Radium Institute of the Academy of Sciences of the USSR.

Jakov J. Frenkel from Leningrad pointed to the disintegration of heavy nuclei—almost at the same time as Bohr—in March, 1939, with the help of the 'liquid drop model', and indicated many explanations for an explosive course of the fission process. In June 1940 the Russian researchers Georgi N. Flerov and Konstantin A. Petrshak extended the knowledge of uranium disintegration with a further phenomenon. They were able to report that uranium atoms could also spontaneously disintegrate. This effect did not find any practical applications.

Within a year there appeared on the theme of 'nuclear fission' over a hundred publications. Each new contribution in the USA, France, Germany and the Soviet Union was to be immediately taken up by their colleagues in the field in their various countries, underpinned, interpreted, and built upon with particular investigations. No one had any secrets from one another. Since the announcement of the fission of the uranium atom by Hahn and Strassmann had appeared the international family of atomic researchers had moved yet closer together, or else behind the apparently peaceful, productive scientific exchange of views they hid a keen rivalry. Perhaps they were also afraid that they would come too late to the crucial breakthrough which must lead to the freeing of atomic energy.

11.8 Fantastic Energy

All in good time, ideas queued up about how to make practical use of the fission of uranium. Expectations were awakened in the German public by a report in the widely read popular science periodical *Die Umschau* ('The Review') of 26 March 1939. In it was said that from the 'revolutionary importance' of uranium fission which had been shown by Hahn and Strassman it is 'not improbable' that 'some uses will be derived from this invention by technology'.

From the specialist literature it is well known that the French researchers had adopted the idea of a 'uranium machine', early decided upon in outline. In April 1939 Jean Perrin published his calculations for how a self-sustaining chain reaction would have to run in a sphere of uranium oxide with a diameter of 1.3 metres. Joliot-Curie and his colleagues expectantly experimented with a hollow copper sphere, which they filled with 300 kilo-grammes of uranium oxide. As an emergency measure, a cadmium salt solution was kept ready at hand. In cadmium a substance had been found which caught neutrons and in doing so would force to a standstill a chain reaction once it got running.

It took courage to carry out such experiments, because nobody knew at the time whether the experimental set-up would fly into the air in a split second. A year later the experimental proof was first found that only the uranium 235 obtained by preparation was the actual atomically explosive substance.

It all looked as if there was a bit of academic sport, for in the issue of *Naturwissenschaften* of 9 June 1939 information and ideas which were already very concrete were gathered and published in a many paged review article. As the author of this work, which bore the title 'Can the Energy Content of the Atomic Nucleus Be Made Technologically Useful?'[8], Dr. Siegfried Flügge, assistant in Hahn's institute, was indicated.

From the energy released by the fission Flügge calculated that the energy content of one cubic metre of uranium oxide would be sufficient to lift one cubic kilometre of water, weighing one thousand million tons, to a height of 27 kilometres. This comparison was impressive. The physicist had already written what at that time of the state of discovery was everything that was necessary for setting an energy-producing 'uranium machine' into action.

From America came the surprising announcement that success had been achieved indeed on 27 July 1940 in the technological exploitation of the fission energy of uranium. Had the Americans built the first energy releasing uranium machine? The high expectations were not filled. Nevertheless, the original experiment was mentioned in the periodical *Mechanical Engineering*, and it was announced that "the energy of a disintegrating uranium nucleus has been used 'technologically', with the help of amplified ionisation impulses from a disintegration fragment, to set in operation the new transmitter of the Boston Radio Station".

Also other announcements were, in spite of their often speculative character, of a generally unbroken optimism, with uranium fission now finally the key to finding an inextinguishable source of atomic energy. Looked at in the cold light of day, nothing useable was actually held in the hands. The situation resembled a race in which only the goal was well known. After the starter's pistol everybody ran hither and thither, for nobody knew the way to the goal. A guiding hand was lacking. The experimentalists thought that now it would be that only once the theoreticians had gotten the chance of regaining the ground which they had lost to experimental atomic research since the work of Hahn and Strassman would things be seen to happen.

---

[8] *'Kann der Energiegehalt des Atomkerns technisch nutzbar gemacht werden?'*

For that reason it was gratefully welcomed that Bohr and his former pupil Wheeler published a comprehensive theoretical study of the 'Mechanism of Nuclear Fission'. For the time being it provided the very long overdue explanation for the surprising disintegration of the uranium nucleus and strengthened Bohr's hypothesis that only uranium 235 could be disintegrated by neutrons.

Bohr and Wheeler's treatment appeared in the *Physical Review* of 1 September 1939. It was as fateful date for atomic physics as it was for politics, and especially for the people of the world. On that day, with Hitler's attack on Poland, the Second World War began. Two days later Germany, England and France were at war. From now on a veil of distrust came down upon the previously so freely pursued international atomic research. Almost abruptly the contact of the German scientists with their colleagues abroad was given up. Germany had once again suddenly started a war, and it would have to be feared that this time, too, their science would be misused for the art of warfare, and still more destructive and inhuman weapons would be invented and constructed, like the poison gas of the First World War.

Above all, the scientists who had been driven out of Germany or who had fled from the Nazis thought with horror what it would mean for humanity if this war were to be carried out with atomic weapons, for after the latest discovery that could no longer be excluded. With the outbreak of war their thoughts and actions were dominated by one frightful prospect— Hitler might succeed in possessing the atom bomb.

12

The Threat of Hitler's Atom Bomb

12.1 A Memorable Conversation

Late one evening in February 1939 the doorbell rang loudly in the Berlin house of the young philosopher Georg Picht. At the door stood a friend of his youth, the physicist Carl Friedrich von Weizsäcker, who said, out of breath, "I have just come from a conversation with Otto Hahn, who thinks it is probable that we can construct a bomb which, if it is done the right way, could destroy London".

That is how Picht described the moment in which he was the very first outsider to learn that the building of an atomic bomb was possible.

The two friends discussed the matter until three o'clock in the morning. Von Weizsäcker sought to make clear what moral attitude he should take in case he and his colleagues were forced to work on the building of such a weapon.

The young von Weizsäcker, whose father decorated the office of a State Secretary in the Foreign Office, was no national socialist and foresaw political and military consequences for the threatened world.

Nothing more is known about the content of that conversation between Hahn and von Weizsäcker in February 1939. One will not miss the mark in supposing that the two spoke about the consequences of nuclear fission after a chain reaction entered the realm of the possible. Otto Hahn would have welcomed the advice of his young colleague, for von Weizsäcker had then already shown, at the tender age of twenty seven, extraordinary technical abilities which were happily complemented by his talent for sorting out the philosophical and political kernels of problems.

Although in a few historical accounts another view can be found, Otto Hahn and Fritz Strassmann must have already foreseen in good time that their discovery could also have ominous consequences. "At that time Herr Hahn and I recognised the possibility of the production of energy explosively when in 1938 we had carried out nuclear fission", said Fritz Strassmann about the question in an interview twenty years later. "We kept quiet about it for a long time. The discovery of nuclear fission could, self-evidently, not be kept secret".

How otherwise should we also assess Hahn's remark in his letter of 19 December 1938 to Lise Meitner: *More and more we come to the terrible conclusion that our Ra isotope behaves like Ba . . .* ? A further occasion is worth considering. It is known that at that time Otto Hahn was in a psychologically fragile state. In his *Recollections 1901–1945* we read that he was *so mentally exhausted by Edith's illness and by agitation about Lise* that even in the presence of others tears often came without warning. Lastly, scarcely less gruelling were the countless investigations undertaken with Fritz Strassmann, in which the two worked up to the limits of their physical capacity.

All these circumstances bequeathed their inevitable stamp on the sixty year old scientist. Also the thought that Germany with an atomic bomb could enslave the peoples of the world must have driven Otto Hahn into depression. His conversation with von Weizsäcker on that February day of 1939 must hardly have contributed to the quietening of his churning conscience. The prospect of a terrible weapon of annihilation was looming as the consequence of uranium fission. In his paper 'The Responsibility of Science in the Atomic Era'[1] Carl Friedrich von Weizsäcker said, looking back to that time, "At a single stroke, in March 1939, perhaps 200 scientists in all the great countries now knew that probably it would be possible to build atomic bombs, and even machines driven by atomic power. What were they to do?"

12.2   Orders to Report for the Atomic Researchers

In April 1939 a letter from the physicist Georg Joos at Göttingen landed on the desk of Reich's Education Minister Rust. At a colloquium of physicists, so Joos reported, the technological use of nuclear fission and the advantages of an energy producing uranium machine had been discussed.

Rust reacted unusually quickly and entrusted Professor Abraham Esau, the President of the Physical–Technical Reich's Institution and the subject

---

[1] *'Die Verantwortung der Wissenschaft im Atomzeitalter'.*

area leader in the so called Reich's Research Council[2] with the summoning of a secret conference. Esau rounded up all the experts who, in his own opinion, could possibly be such. In prime place was Professor Hahn. But this particular obligation to speak was one that he could not decline. Besides, thought Hahn, he was only a chemist, not a physicist.... . The problem, however, was of a purely physical nature. In short, Hahn did not go but sent the physicist Josef Mattauch as a stand in. Esau's rancour at the disobedient scientist, whom he accused of 'political unreliability', Hahn was from now on to get a taste.

Since January 1939 Professor Mattauch had been a new colleague at the Kaiser Wilhelm Institute for Chemistry. Hahn had proposed to the Viennese physicist that he take over Lise Meitner's position. When Mattauch arrived cheerfully expectant in Dahlem, Otto Hahn gave the new comrade in arms to understand, in an endearing and charming way, that it was not so much himself but much more his famous mass spectrograph which was needed at his institute.... . Mattauch seemed to be disarmed by such honesty and—stayed.

At the first secret session on the uranium problem, which took place in the confines of the Ministry in Unter den Linden in Berlin on 29 April 1939 there was at first massive reproach of the absented Otto Hahn. Hahn was sharply criticised that he had, in any case, published the crucial discovery. The whole world now knows and profits from it. Mattauch rejected the attack, in that from time immemorial scientists had had the duty to submit their results to the public at large for their information, and to the their colleagues in the subject area for checking. Professor Hahn had done nothing other than that.

If one studies the statements which were made at this secret conference in the Reich's Cultural Ministry one gains the conviction that the appropriate authorities were very interested in a speedy application of the uranium problem. With immediate effect all the uranium ore required would be guaranteed from the rich mines in the Bohemian Joachimsthal and its export abroad would be stopped. In addition it was planned to bring together into a research group the leading physicists, who should work single mindedly to bring a 'uranium machine' into being.

Matters got going. But in another place was strong interest shown. The army's high command set up its own uranium research plan. The start of this undertaking triggered a letter here, too. On 24 April 1939 the Hamburg physicist Paul Harteck and his assistant, Wilhelm Groth, wrote to the Reich's War Ministry to urgently "draw attention to the latest development in nuclear physics, which is that it is probably possible to produce an explosive which is many orders of magnitude more effective than presently available". Harteck's letter culminated in a statement which could not have

---

[2] *Reichsforschungsrat.*

failed to make an impression on them: "The country which makes the first use of it will, in comparison with others, possess a superiority which can not be caught up with".

Neither Harteck nor Groth received an answer. But their letter set off a chain reaction in the military. The army's high command gave the Army Weapons Branch[3] the charge of looking after this matter. From there the letter went on to the research division which the physicist Professor Erich Schumann led. He admittedly personally that he did not hold with the matter, but could only say derogatorily, "Keep these atomic detritus[4] away from me!" (When Hahn heard of this he commented, *This language comes easily to atomic fogeys![5]*) In order not to reveal his weakness in the subject Schumann finally sent the letter on to Dr. Kurt Diebner, an army expert on explosives.

The stone had also been set rolling by the military. In the summer of 1939 the Army Weapons Branch equipped an independent department for nuclear physics, with the direction of which Diebner was entrusted. In Gottow, near Kummersdorf, south of Berlin, Diebner found the opportunity of marking out his experimental field in the armed forces' testing ground for explosives and shells.

When, on September 1, the Second World War broke out belligerent Germany alone possessed two—even if rival—groups, which were tackling uranium research on state commission. The military compelled the collaboration on the armed forces' uranium research project of the specialists needed in apparently the simplest of ways. They sent call up papers to the scientists in the very first week of September. These differed from the others sent to thousands only in that the physicists had to report directly to the Army Weapons Branch. By this measure the rival concern under Esau was eliminated. Also Otto Hahn was not able to evade the orders of the military authorities in the long run. Of necessity he took part in many meetings in the Army Weapons Branch, which in September 1939 were held almost non-stop, and received in his institute the high ranking head of the uranium project as well as—Paul Rosbaud.

September 16
Meeting in the High Command. Development of the programme, etc.. Afterwards Harteck in the laboratory... Telephone call from Esau about uranium, wants to see me!
September 17
Visit by and discussion with... Heisenberg, then Rosbaud, who has come back from Oslo...

---

[3] *Heereswaffenamt.*
[4] *Translator's Note:* 'Kacker(s)' is the coarse German word for fæcal matter.
[5] *Translator's Note:* 'Atomknacker', which is a pun on 'atomkacker'.

September 18
Weizsäcker, Geiger, Laue in the laboratory.
September 22
Prof. Jung, Wa Prü[6] *IX here, Rosbaud... here in the evening.*
September 27
Dr. Diebner in the laboratory. Meeting.
September 28
Weizsäcker in the laboratory.
September 28
Rosbaud in the laboratory.

12.3   Heavy Water

Diebner and the Leipzig physicist Bagge, an assistant of Professor Heisenberg, drew up, as instructed, a 'preparatory plan of work for initial attempts for making use of nuclear fission'. Within the framework of this programme all specialists in atomic research were allocated particular tasks. Soon also Heisenberg and von Weizsäcker were obliged to work on theoretical problems. Other researchers, such as Bagge and Bothe, had to attend to experimental tasks.

In accordance with his duties, on 6 December 1939 Werner Heisenberg handed in his report on 'The Possibility of Technological Energy Production from Uranium Fission' at the Army Weapons Branch. The Nobel Prize winner had been successful in his study with notable conclusions which today must be viewed as the first essential concept for a uranium reactor. "The fission process in uranium discovered by Hahn and Strassmann can also be used for the procreation of energy from the data in hand up to now. The surest method for the bringing into being of a machine suitable for doing so consists in the enrichment of the isotope U235... It is, furthermore, the only method of producing an explosive which surpasses the explosive power of the strongest up to now by many powers of ten... ".

In his preparation Heisenberg referred to another variant: "For the production of energy one can, however, also use ordinary uranium with enrichment of the U235 if one combines uranium with another substance which slows down the neutrons without absorbing them. Water is not suitable for this purpose. Against this, from data in hand heavy water and completely pure carbon fulfil these purposes. Very low contamination can make energy production impossible".

Heavy water, D_2O, is a combination of the hydrogen isotope deuterium. In normal water, H_2O, the heavy form is present in only 0.015 percent. In a litre of ordinary water there is therefore something like one drop of

---

[6] *Translator's Note*: *Wa*(ffen) *Prü*(famt), Weapons Testing Branch.

heavy water. In order to produce the heaviest water a lengthy electrolysis process is required, in which considerable quantities of water have to be decomposed.

Therefore, as Heisenberg had estimated that many tons of D_2O would be necessary for a uranium machine—but for the first investigations, however, at least a few hundred kilogrammes—the other variant now appeared enticing, for producing highly pure carbon (graphite) in adequate amounts was hardly a technical problem, and not an insurmountable one.

The physicist Bothe, from the Heidelberg Kaiser Wilhelm Institute, came, after many measurements, to the result that graphite was not suitable. Heisenberg consequently made a revision—heavy water alone was the only usable moderator for a uranium burner. Only later was it found out that Bothe's measurements had simply been wrong.

From the Army Weapons Branch, the attention of which had been drawn to the strategic value of the heavy water, the Nazis sent out emissaries to search for this rare product. There was only one supplier in the world, the huge electrolysis establishment of the Norsk Hydro firm in Ryukan, 120 kilometres west of the Norwegian capital, Oslo. But the Germans had no success in their endeavour to buy the firm's stock—185 kilogrammes of heavy water—and place further orders. Norsk Hydro regretted they were unable to take on the German instructions.

Such a refusal had not been expected. At this point the Germans could not have suspected that in the meanwhile another buyer had run off with the prize. From Joliot-Curie it was found out that it had been the French government which had succeeded in securing the stock of heavy water laid down from the grasp of the Nazis and in transporting it to their own country. When German troops occupied Denmark and Norway, and the only heavy water factory in the world fell into their hands on 4 May 1940 the situation suddenly changed. Norsk Hydro had to commit themselves to increasing production to the greatest possible. The only buyer was now in Germany. In May 1940 Belgium and The Netherlands also surrendered. On 14 June, Paris fell. A little later, France capitulated.

When the German armed forces marched into Belgium they captured the stocks of the largest uranium exports in the world, of the Belgian company Union Minière, which was mined in the African colony on the Congo. Up to then Germany had obtained monthly from this firm a meagre ton of uranium compounds. From now on until the end of the war many thousand tons of this exceptionally important substance were shifted to Germany. The customers were the Auer Gesellschaft in Oranienburg near Berlin and the Degussa works in Frankfurt am Main.

A few days after the surrender of Paris Professor Schumann and Dr. Diebner from the Army Weapons Branch went to see the Joliot-Curies' institute and confiscated the half-finished cyclotron for their own atomic research. Otto Hahn had turned down an offer to take part in this inspection of the Parisian research site. Also when they were later invited there for a working visit he and Strassmann later made no use of it. The visit to Paris brought nothing to their all too great needs, for Joliot-Curie and his colleagues had succeeded in destroying important documents. Also the French had been able to get the costly heavy water to England in time in an audacious operation. The French atomic researcher himself remained in Paris and joined the resistance.

Despite the failure in Paris German atomic research had achieved a remarkable starting position. As the English historian David Irving put it, the Germans at this time came up with an alarming lead in 'the race' to build an atomic bomb. They possessed the only heavy water factory in the world and a cyclotron soon ready for use, they had brought thousands of tons of uranium raw material into their possession, and were able to tell a whole tribe of scientists and specialists, who were pressing ahead under orders with militarily oriented atomic research, what to do.

In order to be able to speed up the uranium work to the level wanted, the Army Weapons Branch took over the Kaiser Wilhelm Institute for Physics in Berlin–Dahlem without more ado. The Director of the institute, the Dutchman Debye, had put before him the choice of either resigning or taking German nationality. That was necessary because a secret project was to be undertaken in this institute.

Debye went to the USA at the beginning of 1940 to give lectures and never returned. Von Laue was the next possible person for the office of Institute Director. Without taking any notice of the protests of the Kaiser Wilhelm Institute, the Army Weapons Branch installed Diebner as the acting Director, whereupon Wirtz, who himself worked at the Institute, whispered to his colleague von Weizsäcker that now they had "Nazis in the Institute". What on earth could be done about it?

Wirtz was regarded as being full of ideas. Not for nothing had he been entrusted with the project of building the first uranium atomic pile in Berlin–Dahlem. Behind a wooden wall, a building with the deterring and deceptive name "Virus House" had been erected as the location for the future "Uranium Machine".

At the suggestion of Wirtz they had gotten Heisenberg as scientific advisor at the deserted institute in order to outwit Diebner. The latter was soon sullenly on his way back to Gottow, and Heisenberg moved up to Director.

A deep rift between the two arose. Neither did it close when Diebner and Heisenberg were compelled to work together on the secret project. In practice this rivalry at times took grotesque forms which communicated themselves to the individual groups at work. Once, when Heisenberg sought a ton of uranium oxide from the Army Weapons Branch, Diebner blocked it; also Harteck needed many hundreds of kilogrammes. Eventually Heisenberg obtained part of it, but on the condition that Harteck's request was satisfied. But Heisenberg sat on his trove of uranium like the dragon Fafnir on the Nibelungen's hoard. At least, that was how the wagging tongues put it. The Army Weapons Branch had to be called upon, and then only

50 kilogrammes were reluctantly parted with by the Kaiser Wilhelm Institute for Physics. In an accompanying letter Harteck was asked not to contaminate the uranium oxide because Heisenberg needed it for further investigations.

12.4   An Outsider

In parallel with the efforts to get a 'uranium machine', nowadays called a uranium reactor, running, work was started on the enrichment of the fissionable bomb material uranium 235. Wilhelm Groth, a colleague of Harteck, believed that this problem could best be solved with an ultra-centrifuge. Other researchers, like Korsching, sought to enrich gaseous uranium hexafluoride with the help of a diffusion process. Bagge swore by his secret patent, the so called isotope sluice.

These ideas and plans were extended by the original idea of an outsider. The thirty three year old Baron Manfred von Ardenne, who for years had surprised the experts with sensational inventions in the areas of radio, television, and electron optics, had tried his hand at this problem since the beginning of 1940 just as seriously. With powerful plasma ion sources he wanted to separate the much desired uranium 235 in an electromagnetic way.

Von Ardenne had an excellently equipped private laboratory in Berlin–Lichterfeld-Ost. He had also sent a letter to a Ministry at the end of 1939 about "making use of the enormous significance of the Hahn and Strassmann discovery". His communication was directed to the Reich's Minister of Posts[7], Wilhelm Ohnesorge. For more than two years von Ardenne had been under contract to the German Reichspost research establishment. What was more direct for him to do than to turn now to his client? Ohnesorge financed the construction of the 1,000,000 volt Van de Graaff generator von Ardenne wanted, which was able to be brought into operation in 1941. The prospect of further support was held out. Hitler's Minister of Posts had derived pleasure from the atomic problem on completely unexpected grounds.

When the established atomic researchers heard of the ambitions of the individualistic inventor they made contact with him. And so it happened that von Ardenne met Otto Hahn and Werner Heisenberg in Dahlem and Lichterfeld-Ost. To von Ardenne's question of how much uranium 235 would be needed for a momentary chain reaction to run, the two replied in unison, "a few kilogrammes".

Von Ardenne thought it over briefly and then gave his views, very sure of himself. It would be technically possible to separate such an amount of

---

[7] *'Reichspostminister'*.

uranium 235 with the help of his plasma ion source and a magnetic mass separator. It would be a prerequisite that the large electric companies were interested in it. Hahn and Heisenberg looked at each other meaningfully. Perhaps they secretly feared that von Ardenne would understand how to master this technical problem faster than others would like to be possible. At any rate, they showed little enthusiasm and answered evasively. "Fortunately it did not come to joint action in this direction", von Ardenne later said, just as relieved.

The Minister Ohnesorge, who was more impressed with the ideas of his protégé than the distinguished experts of atomic research, used the very next report to the Führer to report to Hitler of the atom bomb, which would guarantee the end of the war, and which could be developed within his own department. A Minister of Posts wanted to give lessons in military advice to the "greatest commander of all time"? Ohnesorge won only scorn and derision.

Despite the research project in the Army Weapons Branch, the military leadership remained sceptical about the new wonder weapon. This ought certainly to have changed in 1942 when, after the victory of the Red Army in 1941 at Moscow, the military defeat of the German army on the battlefields of Europe began to loom.

Manfred von Ardenne stuck to his idea. In order to test its usefulness, with funds from the Reichspost he built a pilot plant for magnetic separation. At the same time he started on the construction of a 60 ton cyclotron. He came to an agreement with Otto Hahn that his institute could also use the large installations. It was not to come to that.

Von Ardenne's enthusiasm for atomic research soon became curbed by moral reservations. A decisive moment for his future restraint had to have been a conversation with Max Planck which the two had on 2 February 1940, face to face. "The present military successes should not deceive a scientist who had learned to think critically", said Planck in the course of conversation, and added worriedly, "Unfortunately, very many just deceive themselves".

"They are thinking about the Hahn discovery of uranium nuclear fission?", asked Manfred von Ardenne. "What will the consequences be?"

"The consequences will be unimaginable if this instrument of power gets into the wrong hands", rejoined Planck.

12.5   The Bomb Will Not Explode

Hahn and Heisenberg's reserved attitude towards Manfred von Ardenne's plans for atomic research must be assumed to have had deeper intentions. The same would be true for Planck's intervention. Also a shot in this direction had been aimed by Carl von Weizsäcker.

In September 1940 von Weizsäcker and his father, who was Permanent Secretary in the Foreign Office, together discussed with Manfred von Ardenne the possible consequences of the looming development of atom bombs. This conversation took place in von Weizsäcker's house in Berlin–Tiergarten. A month later, on 10 October 1949, Carl Friedrich von Weizsäcker accepted an invitation to a return visit to Lichterfelde-Ost. In this exchange of views von Weizsäcker delivered an expert report of how much of time remains a mystery, and that von Ardenne should categorise in his memory what he considered to be "errors of historical magnitude". Was it that von Weizsäcker did not intentionally hesitate to use the word 'errors' as a reference to the development of an atom bomb? Von Weizsäcker in fact explained "very precisely" that he himself was now, as also was his colleague Heisenberg, becoming convinced that for a trivial reason atom bombs in practice would not be able to function at all. At high temperatures, at which unquestionably the uranium fission must proceed, the chain reaction must, purely theoretically, come again to a standstill. The bomb just can not explode.

This opinion was supported by one of the leading German atomic theorists, who only recently had provided proof of his ability. Whilst everything was put into the production of atomic energy from uranium 235, this physicist set out 'A Possibility of Energy Production from U238'. So pronounced his report of 17 July 1940 to the Army Weapons Branch.

According to von Weizsäcker's idea, in a running uranium machine the trans-uranic neptunium, of atomic number 93, must form by neutron capture from the non-fissionable uranium 238, hitherto looked upon as useless. From that it was highly probable that a long lived trans-uranium 94 would be formed by radioactive decay, whose isotope of atomic number 239 would be suitable as an atomic explosive, just like uranium 235. Perhaps it would in general be more advantageous instead to concentrate on the easily chemically separatable element 94 than on the physically laborious enrichment and isolation of the rare uranium isotope 235.

Of these trains of thought von Weizsäcker gave away nothing to his colleague in Lichterfelde-Ost. Such a fundamental realisation could not remain hidden for long from his colleagues in the subject. It was only a question of time until other atomic researchers must hit upon the same conclusion.

One year later the Reich's Post Minister knew about the diabolical properties of the element 94 and its advantages over uranium 235. And, indeed, from a secret report of Fritz Houtermans, 'On the question of setting off a Nuclear Chain Reaction', ran the research article of August 1941, in which it says at the end, "I thank Baron Manfred von Ardenne for the stimulus to do this work and for enabling it". The connection was quickly made. Houtermans, who in his early days had already busied himself with theories about atomic chain reactions and with the reaction themselves, as is well known, had found out "why the stars twinkle", and later preferentially occupied himself with this theme. The students and teaching staff of the

Technische Hochschule in Berlin–Charlottenburg who heard Houtermans' inaugural lecture in 1932 were convinced of that.

When the Nazis came to power Houtermans emigrated to the Soviet Union, was persecuted there, and came back to Germany, where the Gestapo incarcerated him. After his release—noteworthy colleagues in his subject had interceded for him—he found, on 1 January 1941, through Max von Laue's mediation, a position in Manfred von Ardenne's private institute. There he put ideas down on paper. When Houtermans, independently of Carl Friedrich von Weizsäcker, found a new practicable way towards the atom bomb on the basis of his own researches, he got caught up in a conflict of conscience. At that perilous juncture he confided in Max von Laue, who gave him the advice, "Dear friend, an invention which one does not want to make, one also does not make". Whether these well meant words were the right ones to calm Houtermans is doubtful, for it is well known that when someone puts a question the questioner often has a spontaneous reaction. Houtermans sent a telegramme to his former colleague Eugene P. Wigner in the USA with a short request which the latter well understood: "Hurry up! We are nearly there". Probably this must have been in the period between December 1941 and the beginning of 1942.

How 'nearly there' they had been in Germany at that time emerged from an interview with Heisenberg: "Actually, from September 1941 we saw an open road to the atom bomb before us". Not only the alarmingly quick progress made by the German atomic researchers must have impelled Houtermans to have taken this spectacular step. On 22 June 1941 the German army invaded the Soviet Union. A few months later, on 7 December 1941, the Japanese attacked the USA's naval base at Pearl Harbor. Only four days later followed Germany's declaration of war on the USA.

The time of the lightning victory was over. In the winter of 1941 the German armed forces came to a halt before Moscow. At this point it was explained to Hitler by the arms trade people that as the German war economy was facing collapse economic measures should put a check on this dilemma. In his capacity as Chief of the Research Department of the armed forces, Schumann informed all the leaders of the uranium research, on 5 December 1941, that the further pursuit of the work "can only be taken responsibility for if the certainty exists that a (militarily interesting) application can be achieved in a foreseeable time". To that end a conference of the "nuclear physics" working community on the state of the uranium research on 26–27 February 1942 should be called and reported. The Army Weapons Branch and, in turn, the appropriate Reich's Research Council for atomic research prepared this crucial secret discussion.

Two events were planned: short papers for laymen in the 'Home of German research', Berlin–Steglitz, held before the heads of the armed forces, of the NSDAP, and of various departments of the Reich; and detailed subject presentations for specialists in Harnack House, Berlin–Dahlem. In the presence of the Armaments Minister Albert Speer; of Wilhelm Keitel, the

Chief of the High Command of the armed forces; Raeder the Commander in Chief of the navy; Heinrich Himmler the Reichsführer of the SS; and other prominent Nazi leaders such as Hermann Göring and Martin Bormann, a boost was promised for the atomic project. Special care was taken on the day, as a glance at the results of the presentation, marked 'secret', shows from the choice of attractive, easily understood short talks.

The promising introductory talk by Schumann was called 'Nuclear Physics as a Weapon'. Afterwards Hahn reported on the 'Fission of the Uranium Nucleus', Heisenberg on the 'Foundations for Energy Production from Uranium Fission', and Harteck on 'Heavy Water'. To conclude, a lecture by Esau was planned, 'On the Expansion of the Nuclear Physics Working Community by the Participation of other Departments of the Reich and Industry'.

However, the prominent political and military figures did not appear. The blame for this "omission of historical consequence", as a few historians lament, was on—a secretary of the Reich's Research Council. Inadvertently she switched the programme for the extremely complicated scientific lectures with that for the readily understood talks. Or had the already very well informed Rosbaud lent a hand in some way? As Himmler, Keitel, Göring, and the others received that invitation to the, for them, unintelligible programme of the conference in Harnack House, they cancelled more or the less diplomatically. Of concepts such as 'capture and active cross-sections' or 'neutron slowing distances' they did not begin to know a thing.

Under the title 'Physics and the Defence of the Country', even the daily press—the *Berliner Börsen Zeitung* of 27 Feburary 1942—reported this totally secret meeting and the participants: "Problems of modern physics" were treated which were "of crucial significance to the interests of the defence of the country and of the entire economy".

12.6   Nuclear Physics as a Weapon

Anyhow, a few of the Nazi leadership learned about the results of this conference through internal reports. Paul Joseph Göbbels, the Minister of Propaganda, wrote in his 'War Diary' on 21 March 1942: "A presentation about the latest results of German science will be held for me. The research in the area of atom smashing has progressed so far that in the circumstances their results can be taken advantage of for the conduct of this war. It gives rise to such immensely destructive effects that if the course of the war is longer one can look forward to a few horrors and to a later war... This is German science at its peak, and it is also necessary that we should be first in this field... ".

In April 1942 Reichsmarshall Göring signed a decree which prohibited the development projects which were merely concerned with 'post-war interests'. Without doubt, the project for a uranium machine which would provide power fell into this category. Furthermore, the High Command of the armed forces announced that purely research projects would henceforth only be pursued if they would lead within nine months to applications which would shorten the war. Speer, the Minister for Armaments, alone was given the right to authorise any exceptions.

Colonel General Friedrich Fromm, Commander in Chief of the Army Reserve, who met Speer regularly at lunch, gave the Minister of Armaments a bit of confidential information. Fromm said—at the end of April 1942—that the war could only be won with a completely new weapon. He was in touch with a circle of scientists who were on the road to such a weapon which could destroy a city, and perhaps even the British Isles, with one blow.

Speer was also made aware of the lack of regard of nuclear research from another source. From Albert Vögler, since 1941 the new President of the Kaiser Wilhelm Gesellschaft, a complaint about the inadequate funding from the Reich's Research Department had been lodged. Speer considered whether Göring ought not to be put as the representative at the top of the Reich's Research Department. Hitler agreed. The new post seemed to be so suitable for the taste of Göring, who swiftly got for himself the title 'Reich Marshall for Nuclear Physics'[8] and everywhere began to sort out where his ideas about order would have to be put into effect.

There was that Professor Hahn who had blabbed rashly and publicised the secret of atomic fission.... Also the behaviour of the other atomic researchers was extremely irritating. Göring was enraged by his first sight of the secret files on the uranium research. "Something bad is going to happen", raged the new Reich Marshall for Nuclear Physics, "when it is seen how these researchers have sprayed their results about with a pressure such as if they can no longer hold their bladders... Everyone who has an interest in making applications of it already knows about it, or they have heard about it. In the first place, theses pages which these researchers have published cannot even be read; or am I too stupid. So many formulae swirl around that one can not find one's way through. But for their colleagues in England, France, and America precise information is known about what kind of egg has been hatched by their colleagues in Germany!"

On 4 June 1942, Reichsminister Speer ordered leading scientists of the 'Uranium Club'—amongst others Hahn, Heisenberg, Diebner, Harteck, and Wirtz—to report so that he himself might finally gain a dependable insight into and the prospects of the uranium research. The secret meeting took place in Harnack House at the Kaiser Wilhelm Gesellschaft, which a few leading representatives attended, such as President Vögler and General Secretary Ernst Telschow. From Otto Hahn's diary we gather that also Chief of the Army Weapons Branch, General Wilhelm Leeb, as well as

---

[8] *'Reichsmarshall für Kernphysik'.*

other leaders of the armed forces, such as Colonel General Fromm, Field Marshall General Erhard Milch, and General Admiral Karl Witzell were present.

The mood of the military was touchy. For two weeks the Royal Air Force had bombed German towns incessantly. Because of it the scientists were given to understand that their results would now be needed in order to be able to exact revenge. For good or bad, Heisenberg had to stand against this pretended line of approach when he stepped up to the lectern. The nuclear physicist busied himself immediately with the military aspect of the nuclear fission discovered by Hahn and Strassmann and began to explain how an atom bomb could be built.

The members of the Kaiser Wilhelm Gesellschaft not trusted with the latest discoveries, who up to now knew only a little of a 'uranium machine', were all astonished. Ernst Telschow later explained that the business about the 'bomb' had been completely new not only to him but also to the others, and had been something of a shock.

Afterwards the military put the screws on the scientists. Their probing questions were long to remain an unpleasant memory, especially to Hahn, Heisenberg, and von Weizsäcker. General Field Marshall Milch wanted to know how large an atom bomb must be to be able to raze a city like London to the ground. "Something as big as a pineapple", replied Heisenberg hesitantly, showing the size of the explosive charge with his hand. The army generals suddenly sat up and took notice. Heisenberg had to apply his powers of persuasion to dampen the enthusiasm that was rising; at present the manufacture of such a bomb in Germany was financially impossible. In any case, it can not be gotten ready in six to nine months as demanded. Unusually high materials and financial provisions would be needed.

Hahn and von Weizsäcker joined in chorus with the lament sung by Heisenberg. With the low budget for atomic research, it would never be possible even to think about a working uranium machine, for which the navy had also expressed an interest as power units for U-boats. Speer stopped their observations with a wave of the hand; as Minster for Armaments he was able to obtain any level of funding needed, if the project was worth it.

That was discordant music, Hahn felt within. A glance at von Weizsäcker told him that the latter was in great difficulties after the words of Speer. Von Weizsäcker's embarrassment grew when he was told impatiently by Speer to say how much money would be needed. Quite indecisively, the physicist finally named a sum of 40,000 marks.

"It was such a laughably small amount", remembered Milch, "that Speer looked at me and we both shook our heads at the unworldliness and naïvety of these people". Speer, also, appeared to be "rather disturbed at the insignificance of the demands in a matter of such crucial importance", as he wrote in his memoirs. "I had already imagined a sum of 100 million marks might be appropriate when von Weizsäcker's answer came".

A little later the mass production of another new kind of weapon was favoured, the V1 flying bomb, and later on the long range rocket, the V2. And so the decision against further sufficient financial support for the development of an atomic weapon was taken.

For this decision Speer, the Armaments Minister, took into account the following argument: to his question of whether a successful nuclear explosion would be held under control with absolute certainty, or would be set off as a chain reaction, Heisenberg had been guilty of the latter answer. Speer also mentioned this to Hitler on 23 June 1942. Naturally that was not the only reason why Hitler's Minister of Armaments seemed to be so cool towards the matter of developing atomic weapons. There was hardly any doubt that Hitler had also considered atomic weapons to be out of the question in this war. According to Heisenberg's believable presentation one would need at least three to four years to build an atom bomb. That would be too long for the Nazis. Speer opined, "The war would have to have been decided long before that".

Despite all that, atomic research did not leave empty handed. Speer put a sum of one to two million marks a year at the disposal of the uranium project and approved the building of a large underground bunker for the uranium reactor. But that was all. It was almost as if by this decision the project for a German 'uranium bomb' had now been conclusively shelved. In actuality, that applied only to an atomic explosive device with uranium 235.

In the course of the meeting with Speer in June 1942, the scientists had pointed out the necessity of building a working 'uranium machine'. Should this serve only the satisfaction of scientific inquisitiveness? Was the element 94, which did not originate from a running reactor, the most suitable explosive for a bomb?

Heisenberg admittedly protested that he had said nothing of the element 94 to Speer because the 'Uranium Club' consisted not solely of him, Hahn, and von Weizsäcker, but also of the ambitious Diebner and Harteck. The enthusiastic endeavours of these two to build a critical reactor themselves under the ever increasingly difficult war time conditions spoke eloquently.

FIGURE 12.1. Hiroshima a few days after the explosion of the atom bomb, August 1945.

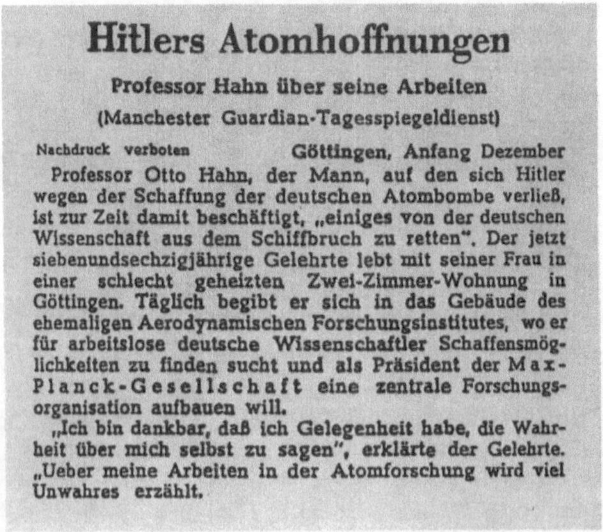

Hitlers Atomhoffnungen

Professor Hahn über seine Arbeiten

(Manchester Guardian-Tagesspiegeldienst)

Nachdruck verboten   Göttingen, Anfang Dezember

Professor Otto Hahn, der Mann, auf den sich Hitler wegen der Schaffung der deutschen Atombombe verließ, ist zur Zeit damit beschäftigt, „einiges von der deutschen Wissenschaft aus dem Schiffbruch zu retten". Der jetzt siebenundsechzigjährige Gelehrte lebt mit seiner Frau in einer schlecht geheizten Zwei-Zimmer-Wohnung in Göttingen. Täglich begibt er sich in das Gebäude des ehemaligen Aerodynamischen Forschungsinstitutes, wo er für arbeitslose deutsche Wissenschaftler Schaffensmöglichkeiten zu finden sucht und als Präsident der Max-Planck-Gesellschaft eine zentrale Forschungsorganisation aufbauen will.

„Ich bin dankbar, daß ich Gelegenheit habe, die Wahrheit über mich selbst zu sagen", erklärte der Gelehrte. „Ueber meine Arbeiten in der Atomforschung wird viel Unwahres erzählt.

FIGURE 12.2. Cutting from *Der Tagesspiegel*, Berlin, 4 December 1946. Translation below.

Hitler's Atomic Hopes

Professor Hahn on His Work

(Manchester Guardian–Tagesspiegel Service)

Not to be reproduced   Göttingen, Beginning of December

Professor Otto Hahn, the man on whom Hitler depended for the creation of the German atomic bomb, is presently occupied with "rescueing something of German science from the shipwreck". The now seventy year old academic lives with his wife in a badly heated two room dwelling in Göttingen. Every day he makes his way to the building of the former Aerodynamics Research Institute, where he seeks openings for work for unemployed German scientists, and as President of the Max Planck Gesellschaft wishes to build a central research organisation.

"I am grateful that I have the opportunity to tell the truth about myself", said the academic. "Many untrue things were said about my work in atomic research".

FIGURE 12.3. The award ceremony for the Nobel Prize Winner Otto Hahn in Stockholm, 10 December 1946.

FIGURE 12.4. *Above:* Otto Hahn... has found the match that lights! *Below:* Warning about the misuse of scientific knowledge.

PROF. DR. OTTO HAHN

COBALT 60

GEFAHR ODER SEGEN

FÜR DIE MENSCHHEIT

VERLAG MUSTERSCHMIDT

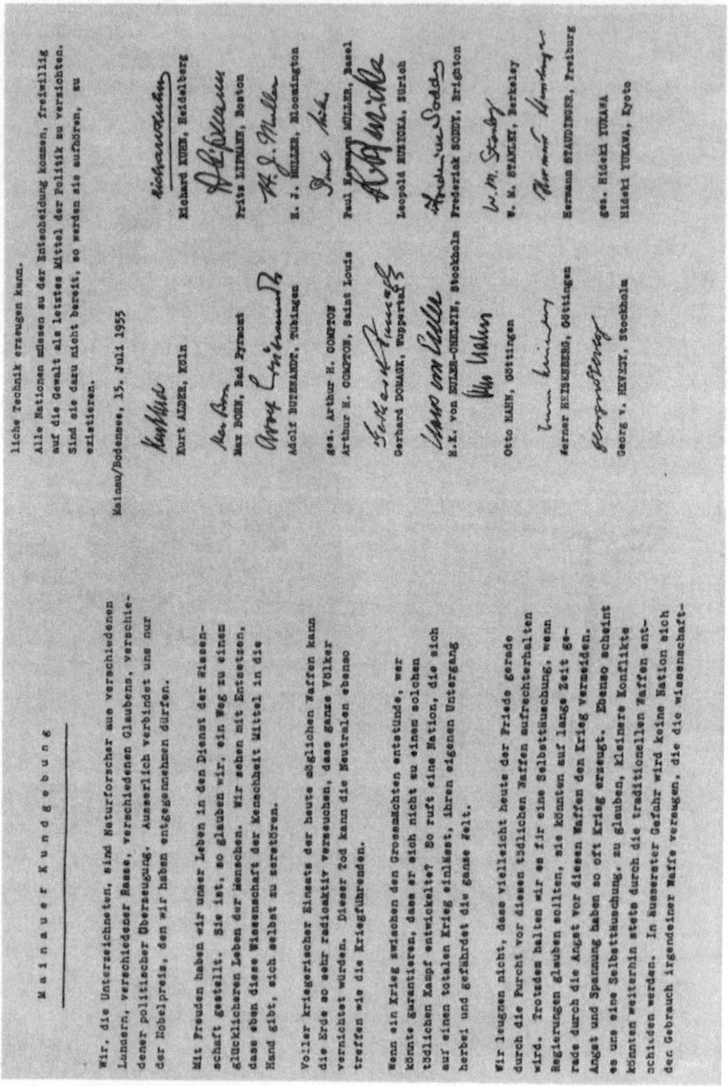

Mainauer Kundgebung

Wir, die Unterzeichneten, sind Naturforscher aus verschiedenen Ländern, verschiedener Rasse, verschiedenen Glaubens, verschiedener politischer Überzeugung. Äußerlich verbindet uns nur der Nobelpreis, den wir haben entgegennehmen dürfen.

Mit Freuden haben wir unser Leben in den Dienst der Wissenschaft gestellt. Sie ist, so glauben wir, ein Weg zu einem glücklicheren Leben der Menschen. Wir sehen mit Entsetzen, daß eben diese Wissenschaft der Menschheit Mittel in die Hand gibt, sich selbst zu zerstören.

Voller kriegerischer Einsatz der heute möglichen Waffen kann die Erde so sehr radioaktiv verseuchen, daß ganze Völker vernichtet würden. Dieser Tod kann die Neutralen ebenso treffen wie die Kriegführenden.

Wenn ein Krieg zwischen den Großmächten entstünde, wer könnte garantieren, daß er sich nicht zu einem solchen tödlichen Kampf entwickelte? So ruft eine Nation, die sich auf einen totalen Krieg einläßt, ihren eigenen Untergang herbei und gefährdet die ganze Welt.

Wir leugnen nicht, daß vielleicht heute der Friede gerade durch die Furcht vor diesen tödlichen Waffen aufrechterhalten wird. Trotzdem halten wir es für eine Selbsttäuschung, wenn Regierungen glauben sollten, sie könnten auf lange Zeit gerade durch die Angst vor diesen Waffen den Krieg vermeiden. Angst und Spannung haben so oft Krieg erzeugt. Ebenso scheint es uns eine Selbsttäuschung, zu glauben, kleinere Konflikte könnten weiterhin stets durch die traditionellen Waffen entschieden werden. In äußerster Gefahr wird keine Nation sich der Gebrauch irgendeiner Waffe versagen, die die wissenschaft-

liche Technik erzeugen kann.

Alle Nationen müssen zu der Entscheidung kommen, freiwillig auf die Gewalt als letztes Mittel der Politik zu verzichten. Sind sie dazu nicht bereit, so werden sie aufhören, zu existieren.

Mainau/Bodensee, 15. Juli 1955

Kurt ALDER, Köln

Max BORN, Bad Pyrmont

Adolf BUTENANDT, Tübingen

ges. Arthur H. COMPTON, Saint Louis

Gerard DOMAGK, Wuppertal

H.K. von EULER-CHELPIN, Stockholm

Otto HAHN, Göttingen

Werner HEISENBERG, Göttingen

Georg v. HEVESY, Stockholm

Richard KUHN, Heidelberg

Fritz LIPMANN, Boston

H.J. MULLER, Bloomington

Paul Hermann MÜLLER, Basel

Leopold RUZICKA, Zürich

Frederick SODDY, Brighton

W.M. STANLEY, Berkeley

Hermann STAUDINGER, Freiburg

ges. Hideki YUKAWA, Kyoto

FIGURE 12.5. The Mainau Declaration.

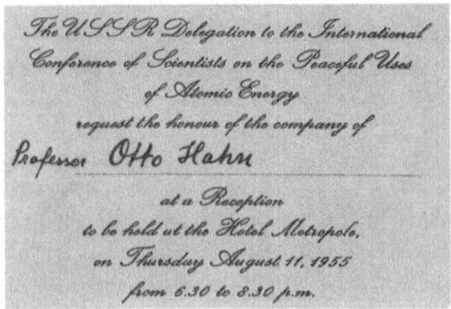

FIGURE 12.6. *Above:* Invitation for Otto Hahn, Geneva, 11 August 1955. *Centre:* Under the chairmanship of the Minister, Herr Strauss, the German Atomic Commission is constituted in Bad Godesberg on 26 January 1956. From left to right: Werner Heisenberg, Otto Haxel, Otto Hahn, and Franz Josef Strauss. *Below:* Otto Hahn, President of the Max Planck Gesellschaft, in a witty exchange with Bundes President Theodor Heuss, June 1956.

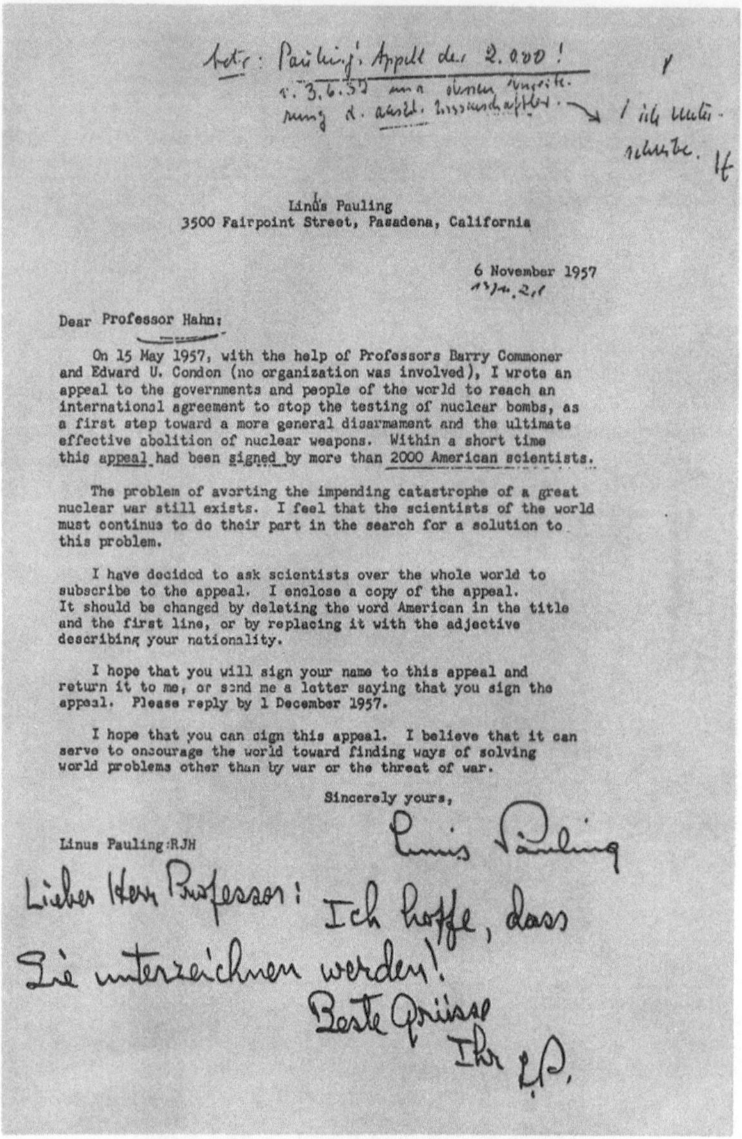

Linus Pauling
3500 Fairpoint Street, Pasadena, California

6 November 1957

Dear Professor Hahn:

On 15 May 1957, with the help of Professors Barry Commoner and Edward U. Condon (no organization was involved), I wrote an appeal to the governments and people of the world to reach an international agreement to stop the testing of nuclear bombs, as a first step toward a more general disarmament and the ultimate effective abolition of nuclear weapons. Within a short time this appeal had been signed by more than 2000 American scientists.

The problem of averting the impending catastrophe of a great nuclear war still exists. I feel that the scientists of the world must continue to do their part in the search for a solution to this problem.

I have decided to ask scientists over the whole world to subscribe to the appeal. I enclose a copy of the appeal. It should be changed by deleting the word American in the title and the first line, or by replacing it with the adjective describing your nationality.

I hope that you will sign your name to this appeal and return it to me, or send me a letter saying that you sign the appeal. Please reply by 1 December 1957.

I hope that you can sign this appeal. I believe that it can serve to encourage the world toward finding ways of solving world problems other than by war or the threat of war.

Sincerely yours,

Linus Pauling

Linus Pauling:RJH

Lieber Herr Professor: Ich hoffe, dass Sie unterzeichnen werden!
Beste Grüsse
Ihr L.P.

FIGURE 12.7. Linus Pauling's Letter to Otto Hahn.

FIGURE 12.8. *Left:* Otto Hahn's spontaneous reaction to Linus Pauling's call for the immediate cessation of nuclear weapon tests. "I subscribe to it". *Above:* 'NS Otto Hahn', the first nuclear powered German freighter run by a reactor in June 1964. *Below left:* Otto Hahn with Frédéric Joliot-Curie at the Nobel Prize Winners' meeting in Lindau, 1 July 1958. *Below right:* The *Berlin Morgenpost*, 27 November 1990.

Gorbatschow erhielt Friedenspreis

BM/dpa Berlin, 27. Nov.

Dem sowjetischen Staatspräsidenten Michail Gorbatschow ist die Otto-Hahn-Friedensmedaille der Deutschen Gesellschaft für die Vereinten Nationen verliehen worden. Die Auszeichnung ehre seine Verdienste „um Frieden und Völkerverständigung, insbesondere um die atomare Abrüstung der Großmächte und die grundlegende politische Neuordnung Europas", heißt es in der Würdigung.

Gorbatschow ist nach dem früheren italienischen Präsidenten Sandro Pertini der zweite Träger des 1988 von der Otto-Hahn-Stiftung ausgelobten Preises.

Der Schweizer Schriftsteller Friedrich Dürrenmatt sagte in seiner Laudatio, Gorbatschow habe den kalten Krieg mit „furchtloser Vernunft" beendet, indem er von seinem Plan der Perestroika nicht abgewichen sei und die Staaten des Warschauer Paktes eigene Systeme und eine eigene Ideologie wählen ließ.

FIGURE 12.9. The gravestone in Göttingen's cemetry.

13

The American Super-Explosive U235

13.1   Press Censorship

Carl Friedrich von Weizsäcker was fond of reading the latest issue of *Physical Review* in the Berlin underground. He continued doing so during the war without worrying about the disapproving looks of the other passengers at the 'enemy literature'. Its reading was worth all that. Almost every issue of this American research journal contained papers about uranium fission. Von Weizsäcker was even more astonished when he found within the issue for July/August 1940 nothing other than harmless articles on the uranium problem.

When he drew Otto Hahn's attention to this strange state of affairs the latter said only one word, "censorship". The Americans must have also recognised the strategic value of uranium for the current war. It was to be feared that they, like the German atomic researchers, were working on the use of uranium fission for warfare. It was self-evident that German scientists at this time similarly had drastic restrictions on publication. As usual, Hahn did not stop because, as he always emphasised, his publications were always only of a general nature, and he must publish something, otherwise, quite correctly, the other side would believe that his institute was now working on secret projects. For reports on work which were to be delivered to the Army Weapons Branch and the Reich's Research Department there existed, in contrast, strict instructions on the observance of secrecy. The daily papers were instructed no longer to publish any form of speculation about the exploitation of atomic energy.

It is quite astonishing that in a research journal, *Nitrocellulose* of November 1940, the following may be read: "One is entitled to great military hopes for the phenomenon of atomic decay, and it ought to be accepted that one day someone will succeed in applying this energy to annihilating complete fortresses with heat, and also melting tanks and bunkers in the blink of an eye".

With this article, entitled 'The American Super-Explosive U235", the journal actually only wanted to "subject to mockery the unbridled fantasy of 'war time inventors'" who always develop their ideas "the most luxuriantly in lands where they themselves do not fight", namely in the USA. That in doing this one might prove to be doing oneself a disservice was never guessed.

The occasion of this article had been a United Press report. UP based it on an interview with Alfred O. Nier of the University of Minnesota in Minneapolis. In March 1940 the American physicist had separated a minute quantity of uranium 235 using a mass spectroscope, and had demonstrated the fissionability of this rare isotope. In this article Nier explained that if one were successful in pressing ahead with the production of uranium 235 in large quantities, "since there appear to be no reasons for not doing so", then a revolution would be near at hand in the field of explosives technology. At any rate, 500 grammes of uranium 235 would equate to the energy content of 12 million litres of petrol. Lastly, in order to put the theory to the vital test of conscience, it is intended to unleash the infernal experiment by setting it off remotely somewhere in the Sahara.

How was this UP report to be understood? Was it, to use Hahn's words, once again *pure American exaggeration*? Or did the Americans really intend to produce uranium 235 on an industrial scale?

The *New York Times* on 5 May 1940 put the UP report about uranium 235 on the front page, over seven columns long. The paper worked out for the astonished Americans that 500 grammes of the uranium 235 isolated by Nier would release the same energy as 5,000 tons of coal or 15,000 tons of dynamite. In Germany—it said further on in the press report—they had already begun to exploit this immense explosive effect for the war. Every available scientist and technician had received the order "to set aside all other work" in order to dedicate themselves feverishly to this task in the laboratories of the Kaiser Wilhelm Institute in Berlin.

The author of this article was the science editor of the paper, William Laurence, who had connections with many prominent American atomic physicists. During the course of the year of 1939 Laurence had already written an article about uranium fission, but intentionally avoided mentioning in it the military aspects. The reporter then ran across the path of Professor Debye. From him and other emigrant scientists he then found out that at the Kaiser Wilhelm Institute for Physics secret work on uranium was to be pursued. It seemed to Laurence that the time for action had come.

The journalist published his article in the *The New York Times* in order to induce Washington finally to take the initiative itself in uranium research. But nothing happened. As far as Laurence could establish, the only 'success' his article had in his own country was that a senator in California began publicly to denounce atomic energy as a rival of the petroleum industry. In the foreign press there were only mocking comments about the new American 'super-explosive'. Otherwise it appeared as if America had slept through the start of the atomic era....

13.2   An Historic Letter

Although Laurence appeared to be incorrectly informed about this matter, if he is to be believed he was the first to have made the US Government aware of the dangerous consequences of uranium fission. For fear of a Hitler atom bomb a desperate step had been taken by a few emigrants. The names of three scientists, Einstein, Szilard, and Wigner, made history in this connection. Leo Szilard was a resolute opponent of fascism and had to leave Germany in 1933 because he was a Jew. Eugene P. Wigner was forced to give up his position as privatdozent at the Technische Hochschule in Berlin–Charlottenburg. In July 1939 Szilard and Wigner sought out their famous colleague in order to speak to him about their fears. Einstein seemed to be taken by surprise. He had not thought of the possibility that Hitler might be able to go ahead with an atomic weapon, or even that it could be possible to make one. The situation now seemed to him to be dangerous.

On 2 August 1939 Einstein signed a jointly drawn up letter to President Roosevelt which was both a warning and a demand. In this letter it was said that on the basis of uranium fission which had been discovered it was now possible to make "a new kind of bomb of the highest detonation power. A single bomb of this type transported in a ship and exploded in a harbour could, in those circumstances, completely obliterate the harbour and part of the surrounding region". Germany had already already started measures to drive ahead uranium research.

To the present day, it has been discussed whether this fateful letter written by Einstein was justified. Einstein himself in later years uttered a bitter reproach. "I committed a great mistake in my life when I the signed the letter to President Roosevelt in which I recommended the making of atom bombs. Yet there existed a definite justification—the danger that the Germans would make them".

Einstein's letter, which in the meanwhile has become a historic document, was the prelude to a central American atom bomb project. It was set in motion, although admittedly with bureaucratic slowness, and needed many an impetus—such as two further letters from Einstein (7 March and 25 April 1940)—until it grew into a gigantic enterprise. When the USA itself entered the war and the news out of Germany piled up, which—like

Houtermans' message—proclaimed the lead had by Germany's atomic research, this programme was raised to the highest emergency priority.

The American military took the 'Manhattan Project', as the cover name was known, firmly into their hands. Under the military direction of General Leslie R. Groves, all the important atomic researchers and hundreds of thousands of technicians and workers were brought together. Groves had made a name for himself by building the US War Department's Pentagon in record time. Now he enjoyed the 'special pleasure' of looking after the hitherto largest collection of 'crackpots', as he once sarcastically referred to them. Thus did the General describe scientists of the rank of Bethe, Fermi, Franck, Lawrence, Nier, Robert Oppenheimer, Seaborg, Szilard, Teller, Harold Urey, and Wigner. Later were added Frisch and Peierls, and for a while Bohr, who in 1943 was able to escape the clutches of the Gestapo by a hazardous flight.

Around this 'bunch of jokers' and their secret activities General Groves drew a steel net of counter-intelligence with the help of the American Secret Service. In this way the work on the American atomic bomb—according to Groves the 'best kept secret of the Second World War'—was to be completely hidden from the eyes of the public.

Otto Robert Frisch, Lise Meitner's nephew, who together with her provided the first theoretical explanation of uranium fission, was caught unawares during a visit to Great Britain at the outbreak of the war. He remained in Birmingham. Together with another emigré, Rudolf Peierls, he worked out in 1940 a plan for the 'construction of a super-bomb'. This report and a further memorandum formed the prelude to a British atomic research project. At the end of 1941 it ran under the cover name of 'Tube Alloys', and was later to be included in the American 'Manhattan Project'.

In the USSR, at about the same time, similar operations were planned. In 1941 Academician Nikolai N. Semeonov referred in a letter to the People's Commissariat for Heavy Industry to the possibility of the production of an atomic weapon. After the German attack plans for atomic research would have to be postponed. One year later, in May 1942, the nuclear physicist Georgi N. Flerov called upon the State Defence Committee "to begin immediately upon the construction of the uranium bomb". At that time the Soviet Union already had at its disposal information from its spy network that atomic weapons were to be worked on in Germany and the USA. Work on its own atomic bomb programme was, however, only able to be undertaken again after the end of the war, under the leadership of the physicist Igor V. Kurchatov.

13.3   The Devil's Business is Done

For the 'Manhattan Project' the US Government roped in the largest groups of companies in the land. In solitary areas towns shot up like mushrooms out of the ground. In Oak Ridge, in the state of Tennessee, factory plants over half a mile long were built for the separation of the uranium isotope by gas diffusion and electromagnetic processes.

On an expanse of 700 square miles large industrial plants were built at Hanford on the Columbia River for the production of the element 94, the importance of which as an atomic explosive the Americans had recognised towards the end of 1941. The first plans for the erection of this gigantic production plant were ready in December 1942, twelve days after that decisive experiment of Fermi. Under the grandstand of a sports stadium in Chicago Fermi had stacked up his atomic pile from six tons of uranium, 36.6 tons of uranium oxide, and 315 tons of the purest graphite, and had successfully 'ignited' it. For the first time ever, on 2 December 1942, a 'uranium machine' produced energy continuously, even if only 200 watts. A self-sustaining chain reaction had been set in action which was capable of being moderated, and therefore in no way explosive. The Americans had succeeded in the crucial breakthrough of using as a moderator that which the German researchers had dismissed as doomed to failure—graphite.

By the spring of 1941 the American scientists under Glenn T. Seaborg had obtained microgramme quantities of the element 94, plutonium, producing it in the laboratory and demonstrating the fissionability of the isotope 239. The atomic industry waited for more results of research in order to be able to separate the plutonium forming in the uranium pile. However, one had first to know the chemical properties of these artificial elements. At the end of 1942 500 microgrammes (half a milligramme) of a purified plutonium salt had been produced—an amount too small even to be able to make a pinhead. Even so, Seaborg and his group determined the properties of this pure trans-uranic with cleverly devised methods of ultra-microchemistry. With this remarkable achievement the Americans created the foundations for the technical production of plutonium in Hanford.

On an impassable high plateau in the midst of the New Mexico desert the USA's actual atom bomb laboratory was set up—at Los Alamos. Here worked prominent American scientists under Robert Oppenheimer, the 'father of the atom bomb', on the final stage, the putting together of the pure plutonium 239 or uranium 235 into an atomic explosive device.

Although later Oppenheimer vouchsafed, filled with remorse, that in Los Alamos he and his colleagues had "done the Devil's business", he found it "technically sweet" to work on the project. When it was known that the bomb was 'feasible', Fermi also was dazzled by this 'beautiful physics'.

Making a bomb is, in principle, easy, but suicidal. In certain doses the fissionable isotopes are harmless. However, as soon as they reach the 'critical mass', which the Americans declared as the most important part of their 'atom bomb secrets', the explosive chain reaction takes place in a split second. It requires only a single neutron as the initial igniter, and which is always available. One therefore requires 'only' to bring two sub-critical masses together very quickly, best done with suitable explosive charges, in

order to set the bomb off. The identification of the critical mass proved to
be a ticklish job. When the American physicists ventured to tackle it, as
they themselves vouched for, they had the feeling that they were tickling
the tail of a fearsome, fire breathing dragon whilst it was asleep. The critical
amount of atomic explosive lay between five and twenty five kilogrammes
of 'bomb pure' plutonium and uranium, respectively. As Heisenberg had
already told the German military, the actual atomic explosive is in fact
"no bigger than a pineapple'.

13.4   The Worries of General Groves

"At the beginning of 1943 we thought, for the first time, that it was possible", the leader of the Manhattan Project justified himself in his *Memoirs*,
"that the German nuclear energy research programme had progressed to
a point at which Germany could use atom bombs against us, or, which
seemed the more probable, against England".

When the western Allies were on the point of preparing for the invasion of
France it was again General Groves who warned the allied operations staff
in March 1944, about radioactively contaminated blockade areas which the
German troops might be able to leave behind.

Yet other worries weighed down upon Groves. Since the end of 1943 the
Nazis' Führer made threats of terrible weapons of revenge. In the presence
of Keitel and the Foreign Minister Joachim von Ribbentrop, on 5 August
1944, Hitler bragged to the Rumanian dictator, Ion Antonescu, about his
'secret weapons', the V1 flying bomb, the long distance V2 rocket, and a
new revenge weapon, the V3, the development of which had progressed to
the experimental stage, and was so massive in its effect that all life within
a radius of three to four kilometres would be wiped out. The leap from
what was then possible to this new kind of explosive substance—so Hitler
explained—was "larger than that from black powder to the explosive common at the beginning of the war". From that and the hints in the German
technical press, the Americans must gather that the Nazis were marching
on unwaveringly towards their goal. For this it is sufficient to hunt down
the official *Progress Reports of the Kaiser Wilhelm Gesellschaft*[1] in the
journal *Die Naturwissenschaften*. It is evident from the yearly report for
1941/42 that in Otto Hahn's institute "the work on uranium fission was
pushed ahead" and 'special investigations' were occupied with the precipitation of and the chemical properties of the element 93 (from which, as is
well known, the atomic explosive plutonium originates).

One year later—in the *Progress Report* for 1942/43—it is stated that in
the Kaiser Wilhelm Institute for Physics under Heisenberg "nuclear physics

---

[1] *'Tätigkeitsberichte der Kaiser-Wilhelm-Gesellschaft'*.

work was continued" and the Institute had been "brought in to work on special questions for a considerable part of the time". Finally one could see in black and white in the daily press, for example in the newspaper *Der Angriff* [2] of 10 November 1944, that Germany had been busy with uranium research. In an article 'Power to raise one thousand million tons 27 km', it says, "German researchers have taken up the idea of now mastering this problem. A new era will come in, the consequences of which can not be predicted today".

In November 1944 a further alarming piece of news fluttered onto Groves' desk. The General candidly confessed that no other report during the war had given him such a terrible fright as that one did. In southern Germany, in the neighbourhood of Hechingen, where, according to the latest reports, the Germans had transferred the atomic research, the Allies' aerial reconnaissance had detected an extremely suspicious object. In an improbably short time there had sprung up a forced labour camp and a complex of industrial plants in the hills, a railway had been built, high voltage power lines erected, and an enormous amount of material had been brought in. Everything pointed to there being a project of the greatest urgency here. Aerial photographs shown to Groves raised the question of whether Germany's Oak Ridge had been set up here, the production site for the enriched atomic explosive.

In order to gain some clarity about the state of the enemy's work on uranium Groves had already, in 1943, brought into being a mobile counter-intelligence unit at the US War Department. They were to prevent information about the German uranium work falling into the hands of the approaching Red Army. In all circumstances the German atomic project would have to be sabotaged.

The so called Alsos Mission—$\alpha\lambda\sigma\sigma\varsigma$ is the Greek word for 'grove'—had unlimited authority, and their goal was to be prosecuted in great secrecy. Colonel Boris Pash was appointed the military commander, whilst the scientific charge was given to the physicist Samuel Goudsmit. Goudsmit came from Holland, and the study of the methods of crime was his hobby. He experienced the great sorrow from the war that his parents met their death in a concentration camp.

Strasbourg was taken by the western Allies after a surprise attack on 15 November 1944. A few days later a man of about forty, in American uniform, rummaged through a pile of papers which the Germans had left behind in the former 'Reich's University'. Outside, on the other side of the Rhine, the sound of battle raged on. But that did not disturb the man. Even nightly air raids did not take him from his tireless searching. Goudsmit was sitting in the office of his colleague Carl Friedrich von Weizsäcker, latterly Extraordinary Professor at the University of Strasbourg.

---

[2] *'The Attack'*.

In the files Goudsmit found notes about the German atomic project. The captured documents, even a torn up letter to Heisenberg which he found in the waste paper basket, conveyed to him an authentic picture of the state of the research work in the summer of 1944. His report to Groves culminated in the statement that Germany was far behind in its atomic research. The new 'super weapon' announced by Hitler could not be an atom bomb.

But Groves did not want to believe the report; the swag had fallen into their laps too easily. He did not trust the torn up letter. Everything could have been 'prepared' and have been a trick by the Germans.

13.5   Otto Hahn's Activities During the War

Which role did the Kaiser Wilhelm Gesellschaft play during this time? "With the outbreak of war the year 1939 brought a great battle for Germany's freedom and future, and also the necessity of certain readjustments for the Kaiser Wilhelm Gesellschaft", it is stated in the official Yearly Report.

In July 1941 the chairman of the board of the largest German steel business, Albert Vögler, was chosen as the new President. The Senate of the Gesellschaft gave him his position because Göring had very personally let it be known that he approved of the choice.

He took up his office, explained Vögler, at a time when the whole of the war also more and more determined scientific work. The first official act of the new President was a directive to all the Kaiser Wilhelm Institutes that all coworkers must be convinced in, even possessed by, faith that the results of their research today guarantee the victory of tomorrow. He expected that they would work "with passionate dedication". That awe-inspiring saying, which has come down from the period of the early 1870s when many industrial firms were founded in Germany, that military strength and science are the two pillars of the State "has now become of deepest importance".

With the degree to which the Nazis had gained the upper hand in the Kaiser Wilhelm Institute, the work for the Institute's members would be complicated with anti-fascist recruitment. Hahn noticed the veiled, and soon quite open, attacks upon his person and his position as Director. *Increasingly I have difficulties because of my lack of membership of the party*, he had to say. His colleague Fritz Strassmann also had to toe the line. He was totally opposed to joining a national socialist organisation. As a result he suffered the longest refusal of a habilitation to which anyone was entitled.

Uranium fission saved the whole situation, Hahn wrote in his memoirs when he described the circumstances under which he and Strassmann had to work during the war years. One followed one's own ideas, so the work on uranium had its good sides because it was seen as important to the war.

It served repeatedly as a welcome pretext. *To the German offices we could correctly say*, thought Hahn looking back, *that it is important to know what 'slag', that is, what atomic species, would be created from the uranium once a machine was got running. For the newly formed kind of atoms absorb neutrons, finally bringing the machine to a standstill, and will have to be separated off from it from time to time.*

One sees that Director Hahn found plausible grounds for the pursuit of the work at his Institute. But were these researches 'important for the war effort', even 'crucial for the war'?

One has to say 'no' to these questions. Professor Hahn and his colleagues busied themselves with the chemical problems of the separation and identification of the fission and transformation products of uranium. Investigations of this kind had to be scientific, and so also had been justified if they were not connected with the military atomic project. With the tasks of the physicists who were trying to build a uranium pile, and who sought methods of separating and enriching uranium 235, with their complicated theories about the mechanism of the chain reaction, Hahn had not the least to do.

Naturally, he could not get out of certain obligations if he was summoned by the Reich's Research Ministry to the routine secret discussions or was called upon to make a written report. Apart from that, Hahn and his colleagues worked very independently, isolating their fission products, and were satisfied when in 1944 they were able to put together in a survey about 100 various isotopes of 25 elements of atomic number 35 to 58. Amongst them were substances the decay time of which was only a few seconds.

The Americans published a detailed table after the war. With the technical capacity and large number of staff, they had all the detailed favourable prerequisites for doing so. Hahn and his coworkers were later very content that the Americans had, nevertheless, not discovered many short lived fission products.

When Otto Hahn heard of the considerable difficulties of his colleagues in building a uranium machine, he also came to doubt whether the *thing* would ever be able to be got working at all. That is also documented in his talk on 'The Transmutation of Chemical Elements'[3], which he gave on 10 February 1942, in Harnack House, and thus, interestingly enough, only a few days before the secret conference of the 'nuclear physics' working group. When Hahn came to speak about the practical application of nuclear energy he did not conceal his scepticism. He was convinced that *it must be left to the future whether the practical application of the enormous energy accompanying the fission of uranium will ever be able to be made available at all.*

---

[3] *'Die Transmutation chemischer Elemente'.*

He voiced this opinion in the autumn of 1943 during an address to the Swedish Academy of Sciences in Stockholm. We can gather from this that Otto Hahn was convinced that the results of the uranium research would no longer be of significance to the course of the war.

At the end of August 1943 the people of Berlin experienced the first heavy bombing raid. It was the prelude to a series of devastating air raids. In the fear that the Reich's capital was threatened with a catastrophe the authorities ordered the evacuation of large numbers of the important Institutions. One of those was the Kaiser Wilhelm Institute for Physics, led by Werner Heisenberg, the centre of atomic research. It was to be transferred to the south German town of Hechingen.

Then came 11 February 1944. This date was to be etched in Otto Hahn's memory. During the night his Dahlem Institute played host to a bomb. The effect was devastating, for the wing of the building in which lay Hahn's Director's office was completely destroyed. Important documents, Hahn's correspondence with Rutherford and other prominent radium researchers, became a victim of the flames.

When constant air raids made carrying on working impossible, Hahn's institute was transferred to southern Germany to the little town of Tailfingen near Hechingen. The Hahn family found accommodation in the house of a textile manufacturer.

After the unsuccessful assassination attempt on Hitler on 20 July 1944, countless arrests were made and death penalties carried out. The Nazi justice system also began proceedings against Max Planck's son, the Permanent Secretary Erwin Planck, and sentenced him to death. Despite Planck's desperate endeavours, it was not possible to halt the execution of the judgement. Hahn did everything he could to console Planck during this difficult time, showing his deep sympathy in letters. It spoke of the depth of Otto Hahn's character that such expressions of sympathy were dangerous to his own life.

In other cases also Hahn took a stand, and openly protested against decisions of the Nazis. When the situation for some coworkers in his institute became dangerous because they were married to Jewish wives, Hahn was called in. He obtained—on the pretext of *work necessary for the war effort* with *poisonous uranium*—the release of a colleague who was to be registered for forced labour as 'unworthy of military service'.

In the same way Hahn protected the wife of his late colleague Rausch von Traubenberg from being taken away to an extermination camp. *Professor von Traubenberg has been occupied with work essential to the war effort in the field of nuclear physics*, Hahn gave as his objection in a letter of 14 February 1944 to the Gestapo. Only the widow, a physicist, could complete the work of von Traubenberg and put his work in order.

Tailfingen was not only something of a temporary measure. With the provision of the necessary means Hahn was able to build an efficient work place and continue single-mindedly with his work, independently and far away from the problems of the nuclear physicists who carried on messing about with their uranium machine. Everything would have instantly changed this situation if Heisenberg and Diebner had actually gotten their 'machine' to run. Hahn would then have also been integrated into a secret project of the greatest urgency by the military. The entire field of work of 'reactor chemistry' would then have come to him, including the investigation of the diabolical plutonium, the bomb explosive substance produced in the pile. Hahn would then—whether he wanted to or not—have had to tackle the real atom bomb chemistry. He could only hope that it would never come to that.

13.6   The Last German War Secret

The rapidly worsening situation of the war had a catastrophic effect upon German industry. Nevertheless, right up to the end large sums were invested in various secret military projects such as the V weapon programme and uranium research. And so the conditions for carrying on the work on the uranium problem were quite hopeless.

A British and Norwegian commando operation, controlled by the Secret Service informed by Rosbaud, had attempted to blow up the factory in Ryukan, and the production of heavy water had been forced to take more time. In November of the same year the Allies had attacked the factory's plant with a carpet of bombs. The military gave up, left the plant in ruins, and was about to transfer everything back to Germany when the ferries with the last stocks of heavy water were sunk. Investigations into starting heavy water production in the Leuna works in Germany did not stop until the end of the Allies' heavy air raids.

A night raid by the Royal Air Force reduced the Degussa works in Frankfurt am Main to ruins. That meant the end of the production of uranium metal in that factory. Now there was only the Auer Works in Berlin–Oranienberg producing the urgently required pure uranium.

Despite this setback, the German atomic researchers worked in their bunker day and night building their uranium pile. Outside howled air raid sirens, bombs went off....

There was the usual squabbling between the competing research groups. Diebner was able to prove that the uranium cube suggested by him for the uranium pile was more advantageous than Heisenberg's slabs. But the Auer Works did not want to fulfil his order; Heisenberg's slabs had a higher priority of urgency.

Things were not going well with the first German uranium reactor when Göring, in the middle of 1943, demanded to be informed by the Reich's Research Council of the position of the uranium research. The 'Reich's Marshall for Nuclear Physics' was interested in whether work might be being done in the United States on a uranium bomb.

"If, also, the (particular) work does not lead to the creation of a practical, useable power generator or explosive", so it says in a report of the Reich's Research Council dated 8 July 1943, for Göring's personal advisors, "there is, then, on the other hand, also the question of security, that the enemy forces will not be able to come up with surprises for us in this field".

This self-confidence did not last for long. At the beginning of July 1944 Göring's adjutant came to Heisenberg agitated. At the Delegation in Lisbon a threat against the German government had been voiced by the United States. Within a few weeks an atom bomb would be dropped on Dresden or another German city if the Government did not seek peace. What did Heisenberg think about this threat? Heisenberg was quite bewildered. Eventually he managed to come to the following 'diplomatic' answer. He held it to be extraordinarily improbable, but not strictly impossible that America already had an atom bomb at its disposal.

In August 1944 a report of the German news agency 'Transozean-Innendienst' of Ribbentrop's Foreign Office brought some embarrassment. In it was that the Swedish newspaper *Stockholms Tidningen* had received the following incredible sensational news from London. "In the United States scientific investigations into a new bomb have been carried out. Uranium serves as the material, and if the power bound up in this element were to be set free then an explosive effect of hitherto unsuspected power would be produced. A five kilogramme bomb could make a hole 1 km deep and 40 km in radius.

One sees that the net which General Groves had cast over the 'best kept secret of the war' seemed to be fairly loose-knit. The Foreign Office did not turn, as would be expected, to the Reich's Research Council to enquire after the truth of this report but to Manfred von Ardenne. When the latter read the news he thought anxiously about the results of Houtermans. Von Ardenne explained that the Swedish report deserved considerable attention, and obtaining further information about it was now of the greatest importance! Strangely enough the *Physikalischen Blätter* published this communication in its August 1944 issue. Nobody—except perhaps Rosbaud—knew how this notice succeeded in getting to the editorial office of this specialist journal. The Editor gave the report the headline 'Just Utopia'.

There was not any doubt that Göring was not satisfied with the progress of the uranium work. In December 1943 he made a spectacular re-allocation of personnel. He put in Walther Gerlach as the new Director of all physics research and transferred to him, with effect from 1 January 1944, the direction of the 'nuclear physics' working community. Gerlach long hesitated over whether he should take this post. When he talked it over in more detail with Hahn, Heisenberg, and Rosbaud, all three advised him to take it.

The offer to Gerlach of the appointment came unexpectedly, for the physicist from München in no way had the necessary references for such an office, and, as he himself put it, "it was not on". Gerlach exploited his position as Authorised Representative of the Reich's Marshall for Nuclear Physics[4] to take freedom of decision in the allocation of financial provision and the right to take advantage of freeing scientists from military service. He also thought about his friends. Hahn received from Gerlach during the time of war "many cheques" for his institute pinned together.

With enthusiasm and personal involvement Gerlach pushed ahead with the uranium machine project. He was also not frightened of making protests to the highest places. So it was that he wrote to Reichsleiter Bormann because Heisenberg, von Laue, and others were to be called up into the Territorial Army, which would cause the uranium project to break down, "It is doubtless known to you", Gerlach protested on 16 December 1944, "that this work is something which has unexpectedly come to a position of importance in deciding the outcome of the war".

In order to reach this goal "quite unexpectedly", Professor Gerlach had picked a new hiding place for the first German uranium pile, inaccessible to the enemy bombing raids, in the sleepy little place of Haigerloch, 15 km from Hechingen, and in an old wine cellar at the foot of a steep cliff. 'Cavern Research Station'[5], it was immediately called on a road sign. Here the uranium machine would be brought into action. At the beginning of March 1945 everything was ready for the critical attempt, which Heisenberg and Wirtz were to lead. To this end the available material had literally been scraped together down to the last speck of dust—1.5 tons of uranium and a meagre 2 tons of heavy water.

In the makeshift illuminated cavern the 'cavern researchers' stared spellbound at the object of their investigation. Slowly the lid with the uranium cube fixed to a chain was lowered into the cylindrical reactor vessel. The precious heavy water was then carefully pumped into the vessel—with a running inspection of the neutron emission which steadily increased, and exceeded, all previously known data. At last, the last of the heavy water was in the reactor. But the pile did not go critical.

Heisenberg made a quick calculation and then announced his verdict. With a further 750 kilogrammes of heavy water the reaction would inevitably be set in motion but with more uranium. Where could this amount be obtained? Gerlach wanted to find out if it could be obtained from Diebner. Anyway, the new Director of the project was confident and announced that within the next six months the first 'chemical reaction' would be able to be put into effect.

An intelligent Government, so Gerlach hinted in conversation in those days, would be able to use this tremendous discovery to negotiate better conditions at the end of the war. Germany now knew something which was

---

[4] *'Bevollmächtigter des Reichsmarschalls für Kernphysik'.*

[5] *'Höhlenforschungsstelle'.*

not known to other nations—the secret of the uranium machine. "Unfortunately we have a Government which is neither clever nor even possesses a sense of responsibility", added Gerlach softly.

14
The Hunt for the Atomic Scientists

14.1   The Alsos Mission Is on Target

The Alsos mission pursued three main goals when it ventured into the European arena of war in February 1945: the arrest of the leading German atomic scientists; the seizure of all the documents and materials related to the uranium research; and the dismantling of all appliances and equipment.

In the meanwhile, the Alsos people knew that the centres of atomic research were located in Hechingen and Haigerloch. In addition they had discovered by espionage that the Germans held their supplies of uranium ore hidden in mines. From there they were taken to the Auer Works in Berlin–Oranienburg, the only remaining uranium manufacturer, for their requirements. Groves was plagued by a new problem, for Hechingen and Haigerloch lay in the part of Germany to be occupied by the French, whilst Stassfurt and Oranienburg lay in the future Soviet zone of occupation. "For me it stood out that the American troops must arrive first at these essential places", Groves explained quite categorically.

In order to demolish the Auer Works in Oranienburg, the Americans did not venture into the area of the Soviet troops. For that, Groves arranged for the regional staff of the western Allies to destroy the works from the air. Six hundred 'Flying Fortresses' unloaded their cargos of bombs over Oranienburg on 15 March 1945, and razed the works to the ground. The civil population must have paid for this action with numerous victims.

With a US infantry division as escort, an Anglo-American commando operation pushed forward to Stassfurt. At this time—in the middle of April

1945—the area lay between the American and Soviet fronts. The Americans seized 1,100 tons of uranium ore in all. A cooperage found in the area worked under pressure day and night to make twenty thousand barrels for the dispatch of the uranium ore. Within three days the loot had been gotten to safety behind their own lines.

A relieved General Groves was able to report to Washington on April 23 that "The capture of this material, the largest amount of the available stocks in Europe, conclusively ruled out every possibility that the Germans might make an atom bomb in this war".

Also the 'German Oak Ridge', which had erected the industrial plants around Hechingen in quick tempo, had, in the meanwhile, proved to be a phantom; it was a refinery for oil shale.

Haigerloch and Hechingen were quickly taken by an American commando raid. On 24 April 1945 the Alsos people broke into the cavern and reached 'the last German secret', the almost critical uranium pile. The first thing that the specialists missed was every kind of radiation protection. Also absent were arrangements for regulation such as Fermi had used in his pile experiments. That itself told the Americans that this reactor had never been able to become critical. Otherwise the Germans had, to be honest, the same difficulties as their American colleagues when their heavy water reactor first began to function in 1944: they could not 'turn it off'. Goudsmit judged that the lack of success of the German atomic physicists in building their experimental reactor as a fortunate circumstance. "For without regulation devices and radiation protection the people involved would probably have suffered a cruel death from an overdose of radiation..." .

Irrespective of the fact of it never functioning, the prototype of the first German reactor was photographed and afterwards dismantled. The hiding place for the uranium cube and the heavy water was wormed out of von Weizsäcker and Wirtz after hours of interrogation with the 'promise' that the Germans would again be able to carry out their investigations under the protection of the Allies.

When, a short time later, the French arrived and took command, to take all the supplies of uranium, those left behind presented to them a piece from the laboratory about the size of a sugar cube. In what had been the location of the uranium pile, nothing more than a heap of rubble remained. The Americans had helped themselves to the entire work and blown up the entire plant before their allies had advanced.

In front of the town hall of the little town of Tailfingen the inhabitants gathered on 24 April 1945 to protest about a pointless order. They had to reinforce anti-tank obstacles and organise resistance against the advancing enemy. Otto Hahn made himself their interlocutor and sought out the mayor. *They ought not to erect the road blocks*, appealed Hahn to him.

"The Führer has ordered resistance to the last", answered the mayor, worried by the uproar.

The Führer can no longer order us, said Hahn fearlessly. *You do not know whether he has not already fled to Austria or gone somewhere else. Save your town and let them praise you. If you put up pointless resistance they will curse you.*

Tank road blocks were not erected and so, against the Führer's wishes, Tailfingen was not defended.

The next day, 25 April 1945, a new chapter in the life of Otto Hahn began. The Alsos team 'seized' Tailfingen and drove with armoured vehicles straight to the Kaiser Wilhelm Institute. The building was surrounded. Storming GIs informed Professor Hahn, who looked thin and ill, that he was under arrest. When he was asked about his secret work and reports about his work he simply said, *I have them all here.*

The detainees were put into an armoured scout vehicle between two armed soldiers. Then the column set off, one tank in the front and a second and more military vehicles in the rear. In Hechingen Hahn came together with the detainees Bagge, Korsching, von Laue, von Weizsäcker, and Wirtz.

"Taking Hahn was easy". With these words Groves dismissed the incident. There were difficulties with Heisenberg, who remained undiscovered as if the earth had swallowed him up. "Heisenberg was to us at the time of the German collapse worth more than ten divisions", said the General, justifying the hunt in his memoirs. It would have to be prevented absolutely that a man like Heisenberg "fell into the hands of the Russians".

But Heisenberg had not taken flight at all, unlike so many others. He had merely hopped onto his bicycle to go to his family at Urfeld. Heisenberg must have suspected what was in store for him. His suitcase was already packed when US soldiers appeared with machine guns levelled . . . .

14.2 'No Fraternisation!'

The dispatch of the captured atomic researchers in the direction of France encountered on the way long military columns of the French. It was a grotesque situation, Hahn reckoned, to see how the 'Allies' each sought finally to get the German scientists out.

Otto Hahn would have liked to have remained in the French zone, for it had become well known that Joliot-Curie would be appointed scientific 'commissar'. The two had become friends during a meeting abroad. The French atomic researcher knew how to interpret Hahn's refusal to come to Paris occupied by the Germans in order to inspect his laboratory. Joliot-Curie returned the favour and took care, after his move to Tailfingen, that Hahn's institute remained unmolested.

After many stops, the captives were quartered in a dilapidated chateau in Versailles outside Paris. Here, in a place bursting with history, they heard,

on 7 May 1945, of the unconditional surrender of Germany in Rheims[1].
Shortly thereafter they brought the detained Heisenberg and Diebner to
them. At last, Harteck and Gerlach completed the circle of ten atomic
researchers whom Groves had wanted: Hahn, Heisenberg, von Laue, von
Weizsäcker, Gerlach, Bagge, Diebner, Harteck, Korsching, and Wirtz. The
scientists were left to feel that they were 'prisoners of war'. Contact with the
population or with soldiers was completely stopped. General Eisenhower
had given the word, "No fraternisation!"

Why, indeed, were they detained? No one suspected the real reason. Hahn
once mockingly asked the duty officer, an Englishman, if they really were
"detained for the pleasure of His Majesty"?[2] The scientists heaved a sigh
of relief when they found out that a decision about their fate had been
taken. The point of view of the English had prevailed in Washington; the
internees had nothing serious to fear.

Only years later was it to become known that the fate of the ten researchers had hung by a single thread. Reginald V. Jones, who as Chief
of the Royal Air Force's scientific counter-intelligence also belonged to an
Anglo-American 'Special Committee for Nuclear Physics Counter-Intelligence', told in 1967 that the German academics had been brought to England on his personal orders. By this measure Jones thwarted the plan of
an American General whose name he did not wish to know. The latter had
intended to remove the 'problem of nuclear energy in post-war Germany'
in his own way, in that he proposed simply to put the atomic researchers
to the firing squad.

On the way to England the internees arrived in Belgium. This time they
were better put up in a hunting lodge built in the English style, Facqueval,
in the forests of the Ardennes, in the neighbourhood of the village of Huy.
A beautiful garden surrounded the house and invited the taking of strolls.
The detachment of guards was spread out for the most part back behind
fencing fortified with barbed wire. Otto Hahn used the time of the stay at
this lovely little spot to write down his *Memoirs 1901–1945*, the manuscript
of which embraced ninety two pages. It was his first attempt at putting into
chronological order the most important stages of his life and work.

On 3 July 1945 their detention in Facqueval ended. They went on farther
to England, to the point of culmination of 'Operation Epsilon'. Without incident the aeroplane landed on a military airfield in the area of Cambridge[3].
On the same day, 3 July 1945, the ten researchers moved into their new
quarters, the country seat Farm Hall.

---

[1]See Note 10 in Translator's Notes on the Text.

[2]See Note 11 in Translator's Notes on the Text.

[3] *Translator's Note:* The airfield was probably RAF Alconbury, a military airfield to this day, and very close to Godmanchester.

14.3 'The Day of the Uranium Bomb', 6 August 1945

After their arrival at Farm Hall, also, they were not allowed to notify their nearest ones. The unknown fate of the families left behind filled all of them with great anxiety. The days were crossed off monotonously in Farm Hall. With colloquia, work in the garden, and sport activities, the internees sought to make the time pass more quickly. A happy placing of people together averted the existing personal differences. Often there were temperamental outbursts and differences of opinion. They often behaved suspiciously to each other, and one reproached another that he had given away all too easily his secret knowledge.

One question recurred again and again: what in the meantime had America and England achieved in the field of atomic research? Bagge did not delay his answer. "I am convinced that they have used the last three months mainly to reproduce our investigations".

Every day they urged the Englishman to give them some prospect that they could write letters. However, they did not know why the relevant authorities in the USA held back this permission. Unexpectedly, on August 5, the news came by telephone. Washington had given the green light in this matter, the Germans might write home, but their letters must be censored. Today we know that Washington was able to grant the authorisation awaited by the scientists with such yearning only after 15 July 1945.

On that very day, the Americans exploded their first atomic bomb in the desert landscape of New Mexico. The date for the experimental explosion eventually had had to be postponed, to the regret of the politicians and military. But the atomic industry of the USA, running at top speed, could not complete the order for the necessary amount of plutonium—six kilogrammes—earlier. In the interim, Germany, for which the bomb had been destined, had long since surrendered.

In New Mexico the first atomic bomb test was successfully carried out, and the new American President, Harry S. Truman, to whom the news had been conveyed in Potsdam, now sought to make Stalin understand the American 'superiority' thus gained. "We shall very probably be very soon in possession of a new weapon, which exceeds all hitherto by a long way in destructive power", said Truman casually, on 24 July 1945, to the leader of the Soviet delegation. Stalin remained unimpressed. "He must not have understood me", said Truman to Churchill, annoyed. When Stalin evaluated the day's conference in the circle of the Soviet delegation, he said of that event, "Kurchatov must be spoken to, so that he speed up the work".

Then came 6 August 1945. Hiroshima was razed to the ground with a uranium explosive charge by the USA's first atomic bomb. The second, a plutonium bomb, fell a few days later on Nagasaki. At that time the USA possessed no more atom bombs. "Is it not one of history's cruel ironies", wrote Bagge in his diary, "that on the morning of 6 August we were allowed to submit the first letters to our loved ones, we, amongst whom was to be found Professor Hahn, the discoverer of uranium fission—on 6.8, on the very day on which the first uranium bomb reduced a Japanese city to ruins... ?"

Everything became instantly clear why they had been 'detained', why all the documents of the German atomic research programme had been confiscated, the plants demolished, and lastly that the experimental atomic pile at Haigerloch had been blown up. The Americans wanted to secure their monopoly of the atomic bomb.

For the same reasons the Americans later also destroyed the cyclotron and atomic research sites of the Japanese. That led to a wave of protests, even amongst the American atomic scientists, who demanded the removal of General Douglas MacArthur, who was responsible. "People who cannot tell the difference between the use of a research machine and the essentially warfare nature of a 16 inch gun have no place in leading positions."

Otto Hahn was "the most friendly of the interned professors", Major Rittner described the German in a note in the records. "He has a very keen sense of humour and plenty of common sense. Hahn is definitely dedicated to England and America. He has been shattered by the news of the dropping of the atom bomb because he feels himself to be responsible for the loss of life of so many people as the result of his discovery". After the first excited discussions Rittner took Hahn to one side to make a proposal to him. Because of the wealth of contradictory press reports and the interest in the interned scientists it was desirable that they provided an objective description of their activities during the war. Hahn conferred with Heisenberg and von Weizsäcker. They needed a day to draw up a suitable text. Eventually an English translation was also published.

This memorandum of 8 August 1945 was the first authentic account of the activities of the German atomic researchers during the Second World War. In the text it was emphasised that "The premises for bringing a bomb into being in the setting of the technical possibilities, which were at Germany's disposal, (did) not obtain. The later work was concentrated on the problem of the machine.... Towards the end of the war this work had progressed so far that setting up an energy releasing apparatus would have required only a short time".

After discussions with some controversy, they also said that the others should be ready to sign this short description. Suddenly there were again differences of opinion. Hahn had been the first to put his name under it, and after him von Laue. Then Gerlach, Heisenberg, von Weizsäcker, Wirtz, Harteck, Bagge, Korsching, and Diebner were to sign in that order. A few refused to do so. How was it that von Laue had signed right next to the top when he had done the least for the uranium work? Diebner played the injured party. Should his name, of all people, be put last under the description?

With a conciliatory proposal Hahn sought to remove the 'problem' to the other side of the world—only Heisenberg, Gerlach, Harteck, and he himself would sign it. But the Major would not agree to that. He insisted that everybody was to sign it. Otherwise it could look as if only the signatories had not worked on the 'bomb'. Finally all ten names stood on the paper in the first order, as planned.

"The Germans knew, in principle, that a bomb could be built", Rosbaud commented in 1959, aiming further explanations and interpretations in this direction, "only they did not know how". At the same time, he found fault that gradually "the impression was forming that the German atomic physicists bore no actual nor any moral guilt for the atomic bomb". And on this question Rosbaud had his own doubts.

14.4   A Difficult Decision

For a good many days Bagge had brooded over a calculation. He wanted to analyse the last Haigerloch experiment again. Finally he succeeded in obtaining the result he had hoped for, that the experiment of March 1945, in spite of the low amount of heavy water and uranium, must have led to an over-critical reactor—if, yes, if, instead of the cylindrical form a spherical shape had been used for the reactor vessel.

Bagge's presentation of the evidence elicited from Hahn a sympathetic smile. He now had another problem. With the first letter from home, which had drawn pleasure, in September he also received a letter from Max Planck dated 25 July 1945, addressed to "Herrn Professor Dr. Otto Hahn, Presently Abroad".

The Kaiser Wilhelm Institute was without a President. Dr. Vögler had put an end to his life himself. At the time Max Planck held the office provisionally. As he wrote, from a poll of opinions there had been only one name discussed in connection with a new candidate, that of Otto Hahn.

The acceptance of this post would have meant for Hahn saying a final farewell to experimental science. For an academic who has spent his life's work in practical research and has made exceptional achievements, it is always a painful decision. By then in his sixty sixth year, during his internment Hahn had indeed busied himself with ideas of how to leave the work to the care of the younger generation. However, he had *never* had any ambition to become President of the Kaiser Wilhelm Gesellschaft. As is written in his life's memoirs, he had hoped that this cup would pass from him. *Apart from my blameless political past there is nothing in me from which it could appear that I am suitable for this office*, he said. Von Laue, Heisenberg, and von Weizsäcker, with whom he ran Farm Hall, used all their powers to persuade him. And so it was that in October he wrote to his wife, Edith, at home, *The thought that I might perhaps become the President of the KWG quite scared me, but I can now not really turn it down, so I would like to* . . . .

Then Otto Hahn came up against his reservations about his work on the uranium project once again. Would this war work not speak against his position as President? He knew how difficult it was to refute prejudice. Only shortly beforehand had he been annoyed by one such questionable report. In the *Daily Express* of 30 October 1945 it stated that Hahn had been 'Hitler's No. 1 atomic expert' and would lend a hand in building British atomic research. Professor Hahn had withheld the atomic secret from Hitler in order to hinder the Nazis obtaining the atomic bomb. Shaking his head, Hahn took hold of a pair of scissors to cut out the article and stuck it in his diary. How could he fight against this mountain of accusations. Here in Farm Hall his hands were tied. Hahn thought it over. Perhaps the office of President of the Kaiser Wilhelm Society would be a suitable platform. Besides, the high standing of this position would make it easier for his request to be successfully entertained that science should serve only for the progress of mankind and not for the destruction of life. Such a challenge of achieving worldwide observance appeared to Hahn, after Hiroshima and Nagasaki, to be the order of the day.

The occasion of the celebrations of the bestowal of the Nobel Prize in Sweden on 10 December 1945, Hahn experienced from afar. It was depressing to have to share this day of success and joy with so much doubt. In the press, one read of that year's Nobel Prize winner for Chemistry, the German, Otto Hahn, being "unfortunately unable to come" to accept the honour in person.

In the circle of the internees, immediately after the announcement of who was to receive the Nobel Prize for 1944, a small celebration had been immediately improvised. Max von Laue gave the eulogy and introduced it with a quotation from Fontane:

> Gaben—wer hätte sie nicht?
> Talente—Spielzeug für Kinder.
> Erst der Ernst macht den Mann,
> erst der Fleiß das Genie.

> Gifts—who would not have them?
> Talents—toys for children!
> Only weight of thought defines the man,
> Only diligence the genius.

"The Nobel speech by Hahn was quite delightful", wrote Lise Meitner to von Laue on 1 September 1946, "and the quotation from Fontane fits Hahn just as if it had been written of *him*. I myself was always so deeply impressed by Hahn's deep seriousness and sense of responsibility, not to speak of his great diligence in his scientific work, particularly because he was sometimes somewhat childlike kindly in human problems".

Back at Farm Hall in December 1945, joy broke out with the longed for news that the internees would at last be able to go back home on 3 January 1946. Surprise was mixed with the jubilation. Why was that date chosen, and why not Christmas. The well versed von Weizsäcker finally found the solution by studying English law. A valid reason must be found within six months to justify a further compulsory stay by foreigners on British territory. The British had found none. But they had taken every advantage of the law that they could.

On the planned date, 3 January, a British military aeroplane brought Otto Hahn and his colleagues back to a Germany devastated by the war.

15

A World Full of Prejudice

15.1 A Protest in Despair

Alswede is a village of scarcely fifteen hundred inhabitants in the neighbourhood of the Westphalian town of Minden. The ten scientists were taken there, and should therefore have been pleased to have been within the 'hospitality' of the British occupying power. The news of the return of Otto Hahn spread like wildfire. Once again on native ground, he and his colleagues were confronted with ridiculous reports about German atomic research. *How childish the rumours seem to one, that we made atom bombs for the Americans, and nonsense like that*, wrote Hahn, annoyed, in his first uncensored letter to his wife, Edith, on 6 January 1946.

The stay in secluded Alswede also brought some good with it for the ten atomic researchers. They were at first free from the embrace of the public. That changed when Hahn, von Laue, and Harteck received permission to take part in the Röntgen celebrations taking place in Hamburg on 23 and 24 January. A memorable anniversary. Exactly fifty years before he had begun the era of atomic research with his discovery of Röntgen rays.

The level of the event admittedly did not correspond to the significance of this anniversary. Hahn was able to follow the last lectures—as he confessed—with the help of caffeine tablets. Reporters present yawned with boredom. When, however, it got around that amongst the honoured guests was the famous Otto Hahn, they became restless. Hahn? — was that not the man who had taken the atom bomb to the Americans? In the first interval the journalists ran off, surrounded Otto Hahn, and bombarded him with questions.

Professor Hahn flatly rejected the idea that he and his colleagues were the inventors of the atom bombs which had destroyed Hiroshima and Nagasaki. *This had been quite impossible, because on the 25th April they had been taken prisoner and the first atomic bomb had fallen on Japan on the 6th August, 1945.* America had in fact spent several years building the bomb, two thousand million dollars, and had built vast industrial plants. In Germany seven and a half million marks had been provided for atomic research, as had recently been reported in the newspapers. How could German researchers have been able to produce atom bombs with that amount of funding, as well as with the country's industry being destroyed?

But the press would not let it go. They wanted to have more precise information about his activities during the war. Hahn parried all questions with a regal silence. *We succeeded in producing a radioactive element with the atomic number 93 which, however, was not stable. We knew that it must yield a heavier element with an atomic number of 94 and atomic weight of 239. However, we did not succeed in producing this substance. The Americans have, by means of their strongly advanced research work, managed to find this element 94 and to produce it on an industrial scale . . . .*

". . . and is this element 94 the atom bomb?", one reporter wanted to know.

"You can regard it so", acknowledged the one enquired of.

The response to the Hahn interview, which several West German newspapers published, was discordant. Hahn's denial was not quite believed. One even accused Hahn of lying. Had not the "famous radiologist Professor Hahn" already protested in March of 1936 that he had produced the elements 94 and 95 in his laboratory and proved it completely? Had that not already been acknowledged as an "historic achievement"? With great patience the accused attempted to explain the actual facts of the matter. Ten years ago one had been mistaken. It had been a 'false' element.

After returning to Alswede Hahn wrote on 28 January to his wife Edith, and son, Hanno. *The business of my 'betrayal' of the atom bomb, connected with the Nobel Prize, I find very annoying. But some people obviously can never be advised of anything . . .*

15.2   Uranium Is Like a Curse

A way will have to be found to be able to counter this insane suspicion effectively. Hahn sought a discussion with Heisenberg and von Weizsäcker in order to find his inner peace once again. With Heisenberg's help, he eventually wrote on 2 February 1946 a four page memorandum, 'The German

Work on Atomic Nuclear Energy'[1]. Otto Hahn attempted to describe the events of the last few years as he saw them. *Germany has possessed neither atom bombs nor plants for the construction of atom bombs. The internment was obviously purely a matter of a security measure. We were not called upon to work for the Allies on atomic energy.*

Owing to then current regulations on censorship, the memorandum could not, however, be published. For that reason Hahn made the text available to the press as internal information, part of it would then be allowed to be published, as then also happened.

Prompted by Hahn, Heisenberg and von Weizsäcker made a great effort to write an analysis in depth of their work during the war. Heisenberg attempted to explain the failure of the German atomic project with the following arguments. "It could not have succeeded for technical reasons", he concluded in his report, 'On the Work on the Technical Exploitation of Atomic Energy in Germany'[2], published in the journal *Die Naturwis-senschaften* of December 1946. "For even in America with its much greater reserves of scientists, technicians, and industrial potential, and an industry working without enemy action, it was only after the end of the war with Germany that they were ready". A relieved Heisenberg established that "the external circumstances" of the German researchers had taken out of their hands the difficult decision of whether or not the atom bomb should have been built.

Such reasoning appeared to be logical, but did not hit the bull's eye, which was only admitted by Heisenberg later, whilst in the interim the facts had become well known. In June 1942, when a decision about the continuation of the atomic project should have been made, the German atomic researchers self-evidently did not know just how long the war would go on.

Von Weizsäcker in his reflections followed Heisenberg's reasoning only hesitatingly. "If we had been given a direct order to build a bomb as scientifically and technically perceived possible, who knows whether we would have been able to carry out this order".

And so during the war the researchers remained "saved from the last tough decision". Doubtless a merit of scientists such as Hahn, Heisenberg, and von Weizsäcker was that they endeavoured "to keep the control of the project in their hands", as Heisenberg put it. But it was a dangerous game to play, for who, in that time of terror, would have been able to vouchsafe that the results of the researchers might not have been taken out of their hands? What would have happened if the Nazis had placed their desperate hope in the atom bomb in the same way as they had put

---

[1] *'Die deutschen Arbeiten über Atomkernergie'.*

[2] *'Über die Arbeiten zur technischen Ausnutzung der Atomenergie in Deutschland'.*

it in their V weapons. Researchers like Diebner would then certainly have found the 'moral courage' to push ahead the building of the bomb just as rapidly as, for example, had Werner von Braun his V weapon programme in Peenemünde.

"Perhaps it might be managed to have the atom bomb ready for use in 1945", pined Speer, the Minister for Armaments, of this illusion in his memoirs. "However, the assumption would have been that at an early stage all technical, personnel, and financial means, some of which would have been for the long range rockets, would have been provided. Also, in this view Peenemünde was not only our greatest, but also our most mistaken project".

It is indisputable that the Second World War sped up the production of the atomic bomb. But it was only at the end of the war that the Americans had two bombs available—no more and no earlier. There is much that has been said about scientific errors and mistakes, which were to be blamed for nuclear fission being discovered only at the end of 1938. As is well known, Hahn did not shrink from self-criticism. Secretly he must have been thankful that Lise Meitner and he were not ready in 1935, or even Fermi in 1934, to have succeeded in this fateful discovery. Without suspecting it they had already produced 'atom bombs' on the microscopic scale. It is not to be thought of what might have happened if nuclear fission had been discovered four years earlier and had been able to be ready for action at the beginning of the Second World War.

From the middle of February 1946 Hahn, Heisenberg, von Laue, and von Weizsäcker lived in Göttingen. The British had ordered that these latter three scientists should take up residence in that old university town. It remained unclear whether they would run an institute within the foreseeable future or would have to occupy themselves scientifically. Atomic research was forbidden them, anyway. The Allied Control Law Number 25 forbade every scientific activity which touched upon military areas of interest.

Hahn's move to Göttingen brought with it no improvement in personal circumstances. Quite the opposite, as must be gathered from his account in his letter to Lise Meitner of 23 March 1946: *Near me live von Laue and Heisenberg, each in his own room. Not exactly very opulent—palliasse bed, no armchair, no sofa, no pictures on the wall. But I am encouraged. I have running water, even if it is not hot. The question of the fate of our institute, of physics, and of chemistry is not yet clarified. So unfortunately I can not go to Tailfingen and visit Edith and the Institute.* His wife had therefore yet longer to live separated from him. Only at the end of July 1946 was her removal to Göttingen approved.

The only consolation during this period was the thought of the Nobel Prize, now to be received in December 1946 after a year's delay. An invitation to Sweden was published. Despite repeated visits to the British authorities, Hahn received no permission for his journey. *Uranium is in-*

deed almost a curse upon me, everything is made more difficult by it, wrote Hahn to his wife, disheartened.

At literally the last moment, when Hahn believed all hope to have been lost, the English allowed the journey for him and his wife. The British civil administration linked it with the condition that Hahn accepted an 'escort' from the British military authorities.

Hamburg, extensively destroyed in the war, was the first stop on his journey. In the guest house of the senate of the Hanseatic city, the new Nobel Prize winner gave a reception for the English and German representatives of the press. *I am grateful to have the opportunity of telling the truth about myself*, explained Hahn. *A great deal that is untrue has been said about my work in atomic research.*

When Hahn then heard the usual provocative questions, he became angry. It had been held, with reproach, said the scientist becoming incensed, that the Nobel Prize was the thirty pieces of silver[3] for having *so to say, offered out of his waistcoat pocket* an atom bomb to the Americans after the surrender. Never had he had the ambition, Hahn passionately asserted, to build atom bombs or to produce radioactive *dust of death*, as not long ago was to be read in a newspaper. *I had no interest whatsoever in a military use. I am a man of science and of peace, and the thought that my work will be able to be used for destruction has often tormented me... .*

15.3   Nobel Prize Winner for Chemistry

In Stockholm there were waiting for Otto Hahn an attaché of the Swedish Foreign Ministry, a crowd of journalists, and—Lise Meitner. The first reunion with her after the war was clouded by discord. Lise poured out reproaches upon her friend, that under no circumstances should he have sent her out of the country at that juncture. Although Lise did not openly voice it to Hahn, others were let know that she found it disappointing that he alone, and not jointly with her, was to receive the Nobel Prize.

In certain respects one can understand Lise Meitner's reaction, but approve of it one can not. Also, on this occasion the decision of the Nobel Committee had not been completely free from inconsistency, as Hahn had noticed. Certainly he saw the problem fundamentally differently from his colleague. The Nobel Prize for Chemistry of 1944 was—word for word—awarded "for the discovery of the fission of heavy atomic nuclei". Uranium fission is exclusively chemical, and had not been proved physically, and, to be precise, had been demonstrated by Otto Hahn and Fritz Strassmann alone. Lise Meitner had repeatedly conceded and emphasised the recognition of the achievement of these two, and that the chemical proof of the

---

[3] *'Der Judaslohn'.*

physics effects of uranium fission could have been carried out by no other research team in the world in 1938.

Individually Otto Hahn and Lise Meitner would have been respectively nominatable for the Nobel Prize for Chemistry in 1915 and 1924, and for Physics in 1937, together even more frequently, for the years 1926, 1934, and 1936. On those occasions a joint award of a Prize would have been just, as it was hardly practicable to separate the pair's team work. But in the radiochemical analytical work in the second half of the year of 1938, which immediately led to the proof of the fission of the nucleus, the absent Lise Meitner had no part at all.

The objection that Frau Meitner had herself first suggested the work in 1934 is not convincing. The search for the resultant products from the neutron irradiation of uranium had been initiated by Fermi and his coworkers. Lise Meitner had merely suggested working on these experiments, which was entirely usual and in no way original. But that is the way things go if researchers publish their experimental results, thus disclosing them, and checking them in another way becomes possible.

Doubtless Meitner and Frisch, but not Hahn's lady colleague alone, had merit in the interpretation of the results obtained by Otto Hahn and Fritz Strassmann with regard to the physical character of the nuclear fission. But they did not gain these laurels, because in January 1939 they were the only ones in the world who were qualified in that area. Rather was it through the unqualified revelation by Hahn of his results that they had a lead in time over others, having been trusted by him as theoretical physicists with the possession of this subject matter. As the subsequent events confirmed, they arrived at the same results.

"Lise Meitner's work was crowned with the Nobel Prize for Otto Hahn". This sentence, wandering like a ghost through the literature since 1981, doubtless has not missed its intended journalistic effect, but is not correct. Once again—the Nobel Prize for Chemistry of 1944 was awarded for the fission of the atomic nucleus of uranium, a discovery which no physicist, including Lise Meitner, had deliberately investigated, because it had not been held to be possible. And had, purely for the sake of argument, Frau Meitner been busy in Berlin–Dahlem at the time of discovery, probably owing to these reservations Hahn–Meitner–Strassmann might well not have found the uranium fission. Otto Hahn said, *I fear that Lischen would have forbidden the fission of uranium to me* if she had remained in Berlin . . . .

The reproaches against Hahn that he alone had been nominated for the Nobel Prize and that he alone had received it are just as mistaken—as if Hahn could have done anything about this decoration finally becoming a reality, after he had been nominated for it for over twenty years! The narrowly defined reasons of the Nobel Committee are certainly worthy of discussion: "For the discovery of the fission of heavy atomic nuclei". Nobody could have said anything against it, if Hahn had been presented the Nobel Prize for his scientific joint work, including his achievement in the discovery of the uranium fission. All his colleagues in his subject would not have begrudged him this honour. But the Nobel Prize exclusively for uranium fission? Ultimately it had proved to be a weakness of formulation that had opened the door to unrest.

The constructive and working interest of Hahn and Strassmann in the discovery can be looked at as being equal. But of course the Nobel Committee took into account in its decision that Hahn had many times been in the arena as a candidate for the Nobel Prize because of his performance beforehand, whereas Strassmann had not been at all.

The prize money amounted to 121,000 Swedish krone. From that Hahn made over 10,000 krone to Fritz Strassmann and an unknown sum to Lise Meitner. She passed on her share to Princeton to the Aid Committee for Atomic Physicists run by Einstein. On Hahn's instruction the Nordiske Kompagnie sent in all seventy two food parcels to relatives, friends, and acquaintances in Germany.

Every year the Nobel prizes are awarded on December 10, the anniversary of the death of Alfred Nobel. The traditional celebrations of great extravagance form a high point in the social life of the Swedish capital. In accordance with an old custom the Prize winners for 1946 also received the gold Nobel medal, presented by the Swedish king Gustav Adolf V.

In his speech of thanks Hahn wove in a personal request. *If I open the newspapers of 1945 I read that I had been in Tennessee in the United States, that I had been kidnapped to Russia, that I had already emigrated to Sweden in 1939, and that everything else which has been said about my activities is so incorrect as even to be romance. In actuality, during the war we carried on with our work and published it. Of that we are glad. But not all are. There are not many outside Germany to whom it is clear under what pressure most people in Germany lived during the last ten years.*

To his celebratory lecture, which Hahn gave on 13 December, he gave the title 'From the Natural Transformation of Uranium to its Artificial Splitting'[4]. He included in it his first great warning of the dangers of atomic armaments. *The energy of physical nuclear reactions has been given into the hands of mankind. Should it be exploited for the furthering of free scientific knowledge, social development and betterment of the conditions of life, or should it be misused for the destruction of those things which mankind has created over thousands of years? The answer should not be difficult.*

---

[4] *'Von der natürlichen Umwandlung des Urans zu seiner künstlichen Zerspaltung'.*

15.4   President of the Max Planck Gesellschaft

Shortly after the end of the war the eighty eight year old Max Planck arrived in Göttingen in one of the Americans' military cars. His notebook with valuable notes was the only thing which he had rescued from his library, but that was at last lost to him. When Otto Hahn called on the Nestor of German science on 1 April 1946, to tell him that he was to take over the Presidency of the Kaiser Wilhelm Gesellschaft, he found Planck in poor health, frail, and in the most needy of conditions. Planck looked happy to give up the burden of the presidency, and said so plainly. For Hahn the acceptance was not easy. Right up to the end he had wrestled with a critical examination of his capabilities. Also later, after taking up the office, Otto Hahn at times became conscious of the limits of his abilities. *If I see the versatility of people like von Weizsäcker, Heisenberg, and even of von Laue*, he wrote in June 1946 to his wife, *I am always distressed by how small the extent of my knowledge is. How could I lecture about Leibniz, Newton, or about natural philosophy? The others can all do that. They read their works, possibly even in the Latin original!*

It was not easy for Hahn to be President of a scientific organisation which was to be abolished under a decision of the Allied Control Commission. *For the last couple of months, while I have been President of the K.W.G., I have aged by just as many years*, he wrote despairingly to Walther Gerlach on 13 July 1946.

By skill in negotiating and doughty staying power Hahn got it agreed that for the time being the Gesellschaft could stay in existence in the British zone of occupation, although the British authorities forbade, as was their right, the use of the name 'Kaiser Wilhelm'.

In September 1946, out of the old Kaiser Wilhelm Gesellschaft was constituted in the British zone the new Max Planck Gesellschaft for the Furthering of the Sciences. Its President got to notice how limited his authority was. For the time being he was forbidden to travel in other zones of occupation or to canvass for his society there. That only changed when the official founding of the Max Planck Gesellschaft took place in February 1948 after the consent of the American Military Government. As a result the institutes lying in the French sphere of influence were able to be included in the society.

The Max Planck Gesellschaft is an association of independent research institutes which belong to neither the State nor to commerce and industry. It pursues scientific research in complete freedom and independence, without restriction of its tasks, and only subject to the law. With these words, so full of promise, President Hahn explained the new statutes at the founding assembly in Göttingen. He set up no mean legacy. As in Kaiser Wilhelm's time, the questions of finance were to become a permanent measure of the society's continuation in existence.

16
Atom Bomb Diplomacy

16.1   A Piece of Good News

Amongst the passengers who, in the evening hours of 23 September 1949, took to the Rhine steamer at Bonn there were to be found illustrious guests. The outing had been arranged by the German Physical Society, which was holding its annual meeting in Bonn. Lise Meitner and Max von Laue numbered amongst the participants, and Otto Hahn was also present recalling his old love of physics.

When the three friends had a lively chat after their return from the trip on the Rhine, they heard other colleagues discussing a piece of news which had just been announced on the radio. In the afternoon, President Truman had made an important statement.

Hahn had an unpleasant feeling, and thoughts about Truman's announcement on 6 August 1945, immediately surfaced.

"Perhaps it will be not that just another atom bomb has been exploded?", was his anxious question. To his astonishment, his suspicion was confirmed. But this time it was a bomb the detonation of which evinced agitation in the White House. The announcement was that an 'atomic explosion' had taken place in the Soviet Union. For the government of the United States, which held up its monopoly of the atomic bomb like a national sacred object and used it to form its global politics, a world had collapsed. Had not their experts asserted to the end that "at least two years" would pass before the Soviet Union could proceed with an atom bomb.

The physicists gathered in Bonn were also taken by surprise when the first journalists wheedled their way in amongst them to seek out their opinions. Otto Hahn's comment was to make headlines. *That is good news*, he said spontaneously. *If the Soviets also have the atom bomb that fact will mean rather a safeguarding of the peace than an increase in the danger of war.*

After the crushing of the fascist power the world had not become peaceful. Admittedly the hot war had ended, but the cold war was intensifying year by year.

In April 1949, at the time of the signing of the NATO military pact by the USA, Great Britain, France, Italy, and a further seven states, President Truman proclaimed, abandoning previous 'reasoning', "I am ready to use the atomic bomb for the peace of the world".

For the Soviet Union there was only one reaction to this power politics, that the American monopoly of the atom bomb must be broken. Kurchatov and his group succeeded in building a uranium pile, which went critical on 25 December 1946; in Europe the first 'uranium machine' was working. Under the leadership of Professor Chlopin Soviet researchers worked on a technology for the industrial production of plutonium with microgrammes of element 94. The construction of a cyclotron began in the same year, 1947, with its completion towards the end. With giant strides, the Soviet atomic scientists began to make up the Americans' lead.

With carefully weighed words the Soviet Foreign Minister, Molotov, stated on the 30th anniversary of the October revolution, "It is well known that the United States bases its politics on the possession of the secret of the atom bomb, although this secret can be seen no longer to exist". When the world's press began to make mysterious prophecies about whether and when the Soviet Union might possess an atom bomb, the strong men of the USA denied all rumours which arose. General Groves mockingly said that the matter amused him, and he asked himself only which fairy tales would be served up to him next.

On 29 August 1949 the USSR exploded its first experimental bomb, more powerful than that which destroyed Hiroshima. As far as the production of atomic weapons was concerned, the Soviet Union had caught up with the United States.

16.2   Big Stick Politics

The very unexpectedly quick loss of the USA's monopoly of the atom bomb triggered a mood of panic in the Pentagon. Culprits were sought who must have betrayed the 'American secret of the atom bomb'—as if there ever had been such a secret! It had haunted in the heads of only certain politicians; for scientists it had never existed. A wild hunt for 'atom spies', the priority being to search amongst communists, was set in motion in the USA, which was to culminate in the execution of Ethel and Julius Rosenberg in June 1953.

With a hysteria that took some beating, the possibility of a world wide debate about atom bombs was brought about. With a cry 'the Russians are coming!', a certain James Forrestal in Washington started out of bed and into the street, and finally ended up in a psychiatric hospital. Similar scenes were repeated and were not worth mentioning, except that this Mr. Forrestal had also been the Secretary of Defense of the United States of America. The incident made headlines in the world's press.

Admittedly, the Soviet news agency TASS had, in the name of the government of the Soviet Union, explained on 25 September 1949 that the USSR held to its view about an unconditional ban on atom bombs. But for the world it was clear that atomic weapons represented just as indispensable an instrument of global politics for the USSR and its sphere of influence as they did for the USA and its allies.

The answer which Truman gave to this on 31 January 1950 ran, "I have given instructions for the development of all atomic weapons to be carried forward, including the so called hydrogen, or super, bomb". This decision set the course for a dangerous atomic arms race, for the announcement of the President of the USA sanctioned the building of the most terrible weapon of destruction which mankind had known hitherto—the thermonuclear hydrogen bomb.

The process which takes place continually in the sun and generates its heat—the fusion of the light elements such as hydrogen and its isotopes deuterium and tritium into helium—this process runs in the hydrogen, or H-, bomb at lightning speed with the greatest of destructive power. To launch the reaction, temperatures are needed of at least 50 to 100 million degrees, which on Earth can only be achieved briefly with a conventional atom bomb as a 'match'.

Up to 1 November 1952 there was in the middle of the Pacific Ocean an idyllic island, covered with palms, called Elugelab. It belonged to the Eniwetok Atoll of the Marshall Islands. But from that day the island of Elugelab has existed no longer. It was erased from the map by the first American thermonuclear explosion. The "Super", as the American technical people called it, yielded an explosive power of three megatons (three million tons) of TNT. That corresponded to the collected power of all the bombs dropped in the Second World War, and around two hundred times the explosive effect of the Hiroshima bomb. But this test was only a 'small' H-bomb. The dreadfulness of this super-bomb is that its size has almost no limit.

Where once there was Elugelab Island there now yawns on the floor of the Pacific a crater one mile in diameter and five hundred feet deep. All the larger of the world's seismological stations registered the shock wave of the explosion of the first earthquake triggered at the hand of man.

In Washington they were convinced, after the successful test, that they had made good the loss of the monopoly of atom bombs, and had won a decisive lead over the USSR. This illusion was to last for only three quarters of a year.

On 12 August 1953 the first Soviet hydrogen bomb exploded. The military and politicians in the USA were shocked when their specialists reported that the Soviet Union now had at their disposal a 'dry' hydrogen bomb with the fuel lithium deuteride. But such a bomb was a competitor, because it was transportable. The USA's 'hydrogen bomb' of November 1952 had been a monster of 65 tons. It resembled a gigantic freezer, for the ignition mixture of deuterium and tritium had to be kept fluid by deep freezing. No warplane had been able to transport this monster and take it to its target.

But in the safes of the Pentagon there lay a further project on the drawing board. The Americans had not previously discussed this 'hyper-weapon' in earnest. Now the last scruples had gone. The biggest stick was taken out of the bag.

On 1 March 1954 the USA exploded their first three stage bomb, also called the fission-fusion-fission bomb. Its detonation effect amounted to at least fifteen megatons of TNT. A fiendishly devised principle lay at the root of its working. An atomic explosive device served as the detonator of the actual hydrogen bomb. The two were wrapped in a coat of ordinary uranium (U238), which itself became fissionable by the operation of the H-bomb explosion. Bombs of more phases opened up inconceivable detonation results, which would be able to reach fifty and more megatons. With such 'hyper-weapons', countries and continents could be laid waste at one blow.

16.3   No More Hiroshimas

The frightful effect of the hydrogen bomb was limited not only to its thousand-fold increase in effect over the atom bomb. It also released radiation of such an intensity as had not previously been known on Earth. It was estimated that one minute after the detonation of a middle-sized H-bomb a radioactivity was produced at the site of the explosion which corresponded to one hundred million tons of radium. Such a dose of radiation is completely fatal for man and beast in the bomb's circle of effect.

When the strongest radiation has subsided dangerous fission products still remain behind, which gently fall slowly out of the stratosphere with the radioactive fallout to the surface of the Earth and contaminate a wide area. Especially insidious are the long living radioactive harmful chemicals like carbon 14, which gets into the biosphere, caesium 137, and above all strontium 90, which gets into the bodies of man and beast with the food, and then accumulates in the bones and inevitably produces cancer.

The American Nobel Prize winner Linus Pauling, who had used all his authority to warn about atomic weapons, gave the following example of the danger of fallout. One teaspoon of strontium 90 distributed evenly amongst all people in the world would lead to their death inside a few years. But each super-bomb, Pauling calculated, threw a thousand times as much into the air with its detonation.

From the studies of an English scientist, the H-bomb explosions carried out up to autumn 1956 had probably led to not less than 50,000 people developing cancer.

In newborn children who had been exposed to radioactive radiation in Hiroshima and Nagasaki physical deformities had formed. What genetic damage might result will only be seen in the future.

In spite of this frightful balance sheet, there were even more voices of scientists who trivialised these threatening dangers. Edward Teller, who had earned the name of 'Father of the Hydrogen Bomb', belonged to those scientists who approved of the Americans' bomb tests. From him came the assessment that the radioactive fallout of a bomb test, when spread out over the Earth, "is as dangerous to a man's health as being thirty grammes overweight". No less misleading was another claim of Teller, that a wrist watch with a luminous dial gives the wearer tenfold stronger radiation than radioactive fallout. Pauling disproved both arguments.

Teller, like Oppenheimer, was one of the students of the German physicist Max Born, a decided opponent of atomic weapons, who stated with some worry in 1957, "It is wonderful to have had such clever and capable students, and yet I wish that they had been less clever than wise. Now, through their cleverness, mankind has got into an almost desperate situation".

Scientists and public from many countries turned against the dangerous game of atomic weapons as the means of political pressure and blackmail, and against the irresponsible experimenting with hydrogen bombs, which threatened the continued existence of mankind,

The French atomic researcher Frédéric Joliot-Curie, President of the World Peace Council and Chairman of the World Federation of Scientists, was one of the most outstanding amongst them. In March 1950 the Permanent Committee of the World Peace Congress met in Stockholm. On the initiative of Joliot-Curie the delegates adopted a call to all men to sign a petition for the banning of nuclear weapons. "We demand an absolute ban on the atom bomb as a weapon of attack and mass destruction", the President of the World Peace Council gave as the reason for this step. "We are of the opinion that that government which is the first to use an atomic weapon against any country will have committed a crime against humanity and will be treated as a war criminal". Joliot-Curie noted with satisfaction the worldwide response to his call. Six hundred million people signed this first Stockholm petition.

A month after his announcement of the petition the French scientist was dismissed from his office as High Commissioner for Atomic Energy. For Joliot-Curie this was a signal to continue the battle. He made an effort to gain the support of colleagues in his subject and personalities whose ethical and moral stance he had come to appreciate. He joined in exchanging ideas with Otto Hahn by letter. If Professor Hahn, as he candidly admitted, was not always able to follow Joliot-Curie, the *honourable convinced adherent of Russian ideology*, as he called him, he always spoke with great respect for the humanitarian commitment of his French colleague. The two were united in their view that they should fight against the arms race and warn of the dangers of nuclear weapons experiments.

17

In Conflict with Conscience and Politics

17.1 Atomic Energy Literature Instead of Crime Novels

As the President of the most important scientific organisation of the Bundes Republic of Germany, founded in September 1949, Otto Hahn inevitably met leading politicians of the new government. The first Bundes Chancellor and CDU Chairman, Konrad Adenauer, received Hahn and Heisenberg in Bonn on 12 December 1949. The problems of the research policy and the financial support of the newly founded Research Council were the official agenda.

But Adenauer had other plans. The Chancellor made an approach that the responsibilities of the Research Council might be to stretch the hitherto forbidden areas of research and test the possibility of getting around Law Number 25 of 1946 of the Allied Control Commission, which included atomic research. And he also was not put off by a new Law Number 22 of the Council of the Allied High Commission of March 1950, which expressly prohibited any research in the BRD in the area of atomic energy.

The Research Council[1], which developed into the German Research Community[2], was allocated the responsibilities of focussing on certain areas in which there was a great need to catch up. The backlog in fundamental nuclear physics resulting from the Allies' prohibition was serious. There-

---

[1] *'Forschungsrat'*.
[2] *'Deutsche Forschungsgemeinschaft'*.

fore, in February 1952, a 'Commission for Atomic Physics' was founded within the Forschungsgemeinschaft, of which Heisenberg was named the Chairman—with the express approval of Bundes Chancellor Adenauer.

In October 1954 the incorporation of the Bundes Republic into the North Atlantic Pact took place. After the sealing of the Paris Treaty, which Chancellor Adenauer signed on 23 October, the statute of occupation for West Germany fell, and with it every restriction on atomic research. Because of pressure from France, Adenauer was certainly obliged to make some kind of statement that aspersions were to be cast without foundation[3]. In its fields the Bundes Republic would expressly renounce the production of atomic, biological, and chemical weapons.

This waiver did not, however, exclude their acquisition, storage, and deployment, nor yet the building of its own atom bombs outside its territory, as political opponents pointed out. It should be pointed out that certain circles in Bonn were willing on no account to resign themselves to the role of 'atomic paupers'.

Adenauer entrusted Heisenberg with leading exploratory talks in Washington about the development of a West German atomic industry. He himself found out about the questions of nuclear energy. At Christmas of 1954 Adenauer was for the first time not wrapped up in crime novels, as was usual, but asked Heisenberg for the most popularised books on atomic research and the exploitation of nuclear energy. The Chancellor also wanted to be able to join in conversation correctly.

The admission of the BRD into NATO laid the way open to establishing its own army. This came about in the Bundes Republic in January 1955 to the accompaniment of numerous protest rallies which gave voice to the displeasure incurred in the major part of the West German people against the threat felt by re-militarisation. As a result, the Adenauer government endeavoured to avoid everything that could stand in the way of the ratification of the Paris Treaty, which was to be decided in the Bundestag on 27 February 1955.

17.2   Cobalt 60

The Church President Martin Niemöller was one of those personalities who raised his voice against re-militarisation and atomic arming. Pastor Niemöller belonged to the presidium of the World Peace Council. In 1954 he wanted the Evangelical Church of his country to encourage an appeal against atomic dangers similar to that Pope Pius XII had pointed to in an Easter message to the Roman Catholic world. The church reacted carefully, and at first sought an 'expert', finally finding him in the person of

---

[3] *'Ehrenerklärung'*.

the Hamburg atomic physicist Pascual Jordan. As an expert he was ready to explain that the Pope had greatly exaggerated the threat to the human race of the H-bomb tests.

But Pastor Niemöller did not resign himself to this indolent posture. When the Max Planck Gesellschaft met in Wiesbaden in June 1954 he arranged to have a conversation with Otto Hahn, which Heisenberg and von Weizsäcker also attended.

"Is it really as serious as the Pope has made out?", Niemöller asked with concern. "In actuality it is much more serious", explained Professor Hahn. "Nowadays it is no longer a problem to build a bomb with which one can destroy all human and more highly organised animal life on the Earth at one blow!".

With a few words Hahn described the newest product of the atomic armaments' madness, the cobalt bomb. The USA and USSR had not yet risked testing this new means of mass destruction in a three stage bomb. Only criminal minds, Hahn emphasised, could have thought up such a piece of devilry. By detonating an H-bomb surrounded with a coat of cobalt, this harmless metal would be turned into radioactive cobalt 60. This isotope emits gamma radiation of an intensity which is able to penetrate walls half a metre thick without any difficulty. Whilst the radioactivity after the explosion of a normal H-bomb dies away relatively quickly, for the cobalt bomb it takes five years before the intensity of radiation has declined to half its original level. The affected area would therefore remain radioactive for years. A few cobalt bombs would suffice for wiping out life on the Earth.

The thought of this frightful weapon left Hahn without peace of mind. He felt himself obligated to enlighten the people about it, for to his way of thinking an appeal had the one purpose that all should know what dangers threatened them. The plan grew in his mind of publishing an article about cobalt 60. In it he wanted to overturn the use of the technology behind the deadly effect of the cobalt bomb into the profitable use of the effect of cobalt 60 in the fight against cancer illnesses.

His colleague Heisenberg also intended to give his views publicly on atomic matters. For a long time the Nordwestdeutsche Rundfunk had prepared for a talk by Heisenberg, which was not transmitted for reasons which were not known. Enquiries to the Nordwestdeutsche Rundfunk piled up, so that finally the Prime Minister of the Bundes State of Lower Saxony, Hinrich Kopf, was called in. The Director General of the Rundfunk now had to be called in. Heisenberg had already wanted to give the talk; however, he could not. Adenauer had requested them "To refrain from every public discussion about the question of atomic energy before the final ratification of the European Treaty".

Kopf, the SPD Prime Minister, went to see Professor Hahn to hear the advice of the President of the Max Planck Gesellschaft on this matter. Hahn took the pressure off himself with bold words. What remained of the much vaunted freedom of science? The ban on speaking evidently indicated that the Chancellor wished the people to remain uncertain about the dangers of an atomic war, in order not politically to endanger the planned re-armament.

The first thing Otto Hahn did was to change an already drafted article about 'cobalt 60' which he had prepared for a pamphlet, and which now seemed to him much too neutral. He had to produce a stirring effect. When Kopf read the revised article he abruptly asked Hahn whether he, in contrast to Heisenberg, had the courage to read out the text on the radio.

"Yes, I am ready".

Kopf did not want to doubt it, but mentioned, with clear allusions, the Chancellor's prohibition. "Can you do it, because ... ?"

"Obviously I can do it", replied Hahn. "What my duty requires of me no man can forbid me".

Warnings from his colleagues, even the reservations of von Weizsäcker, who feared a panic amongst the population, Hahn rejected, his mind made up. *I maintain that information is all for the good.*

On 13 February 1955 the Nordwestdeutsche Rundfunk in Hannover broadcast Hahn's talk on 'Cobalt 60—A Danger or a Blessing for Mankind?'. Radio stations in Denmark, Norway, Austria, and the BBC in London also transmitted the contribution.

Hahn's radio appeal brought not merely the long overdue explanation of special atomic dangers. In his speech he went a decisive step further. He called upon the politicians no longer to solve international problems with force or the threat of force: *A united appeal of all responsible scientists who well knew the dangers of a means of war which threatened the world should succeed in bringing those responsible for big politics to the negotiation table.*

It is to the credit of Otto Hahn that he pointed out the dangerous consequences of power politics with inexorable clarity. That Hahn did not find the desired resonance with the call in the mass media—of the major daily papers the only one to publish the text of his talk was the *Frankfurter Allgemeine Zeitung*—was not surprising. Hahn's personal integrity and the choice of timing immediately before the ratification of the Paris Treaty made the 'Cobalt 60 Appeal' politically heavy. Bundes President Heuss said, "I felt, when I read your pamphlet on the cobalt bomb, in what ethical difficulties you were; was it that what I—Otto Hahn—had discovered has been found to be able to become misused as an instrument of disaster?"

Recognition was also accorded to Professor Hahn by Joliot-Curie, who congratulated him on this step in a letter of 2 March 1955. "Once more you have shown the courage which all scientists should show at the peak of their mission. It seems that we are now in agreement on important points." He, Joliot-Curie, became, as he greeted Hahn, one of those scientists from many lands who, despite their various political persuasions, united together in an appeal.

Already by the end of February, Otto Hahn, with Born, von Weizsäcker, and Heisenberg, had voted for further action. In order to promote the effect of Hahn's radio talk, they agreed to his proposal to direct an appeal to the public of the world jointly with foreign Nobel Prize winners, so that they might be warned of the growing atomic dangers.

The draft was written, checked, and sent to a foreign colleague. Some Nobel Prize winners, like Artturi Virtanen and Robert Robinson, declined their agreement, because the appeal did not match their political views. But the majority declared themselves ready to put their names to such an appeal.

17.3 Declaration on Mainau Island

Lindau is an idyllic little town situated on an island in the middle of Lake Constance. Since 1951 the Nobel Prize winners of the three branches of science—chemistry, physics, and medicine—had met there every year. A nephew of the Swedish king, Graf Lennart von Bernadotte, who had set up in practice on the neighbouring island of Mainau in Lake Constance, took on the honorary patronage.

The public at first received the meeting in Lindau in a subdued manner. Who was interested in these very specialised subject lectures of the laureates and the usual 'tittle tattle' of these *family gatherings*? Yet the Fifth Meeting, which took place in 1955, brought worldwide attention. During the concluding resolution on the island of Mainau Otto Hahn and Graf von Bernadotte read out the German and English text of a declaration, which as the 'Mainau Declaration'[4] was to make history.

In the face of the threatening self-destruction of mankind by a war conducted with nuclear weapons, there was, according to Otto Hahn, one alternative only. *All nations must reach the decision to freely give up force as the last means of politics. Were they not ready to do so, they would cease to exist.*

When Hahn was at the end of speaking he felt the need to come to an understanding with a good friend. He approached Hans Hartmann, whom he had known since 1908. Hahn seemed moved as he began to speak. For the historian his words remained unforgettable. *You have now known me for so long. So tell the people that it is I who am guilty of there being atom bombs and Japan having to suffer so fearfully from them, and that our future is endangered. But I have done only my scientific duty. What the powers in this world have made of it is their responsibility. We researchers also carry a responsibility. Only where we can, we must be attentive to possible frightful consequences, and seek to prevent them.*

The Mainau Appeal of 15 July 1955 was initially signed by eighteen Nobel Prize winners. One year later there was a total of fifty two names,

---

[4] *'Mainauer Kundgebung'*, lit. Mainau Expression.

after further laureates such as the married couple Joliot-Curie, Gustav Hertz, Bertrand Russell, and Cecil F. Powell had put their names to the document at the request of Otto Hahn—not to forget Max von Laue. A few days after the Mainau Declaration he wrote to Otto Hahn, "I have read of your call in the newspaper. Naturally I will sign it".

Independently of the 'Mainauer Kundgebung' the English philosopher and opponent of nuclear weapons, Bertrand Russell, on 9 July 1955, returned to a statement on the danger of atomic weapons, at the suggestion of Albert Einstein who had died in April 1955. This Russell–Einstein Resolution urgently called upon the governments of the world to settle their disputes only by peaceable means, for if a future world war were to be atomic the end of mankind might come about.

To find the Russell–Einstein Resolution or the Lindau Appeal of Nobel Prize winners in the western press required a few efforts. The text was printed in its entirety only in the *Frankfurter Allgemeine Zeitung*. Otto Hahn believed the cause of the small response found by the appeal was that the text was too broadly addressed. Nobody felt themselves to be directly spoken to. With von Weizsäcker, Born, and Heisenberg he was agreed that a further appeal, which would be in the same cause, would have to be more concrete in its intended goals. Besides, it might be better, considered Hahn, only to direct such an appeal to the State in which each lived. It would remain to wait to see how the situation might develop in any particular country.

17.4   International Atomic Energy Conference

In December 1954 the General Assembly of the United Nations took the unanimous decision to call a UNO Conference on the peaceful use of atomic energy in August 1955. Geneva was decided on as the location of the meeting. All eighty four member countries were invited. In a letter to the President of the UNO General Assembly of 17 December 1954 Otto Hahn welcomed this decision. *It is a theme which is of interest to the whole world... My personal opinion is that one should draw together... a number of internationally recognised scientists. I think they should be persons who can be regarded as real experts, but who are also expert in the moral and ethical considerations which lie at the heart of future possible outcomes of the use of atomic energy.*

The suggestion had also been approved by the Soviet Union. With the first atomic power station in the world, the USSR had demonstrated that it attached importance to the peaceful applications of nuclear energy. The Soviet Union invited scientists from every state to an international meeting in Moscow on atomic energy research for peaceful purposes from 1 to 7 July 1955.

At a press conference in Bonn Professor Heisenberg let it be known that the Max Planck Gesellschaft would take up the invitation to the Moscow atomic conference. It had not been decided which scientists would travel to Moscow. Possibly he himself would go.

Heinrich von Brentano, the German Foreign Minister for just two weeks, announced in Berlin on 25 July that caution should be exercised with private invitations from the Soviet Union. The insinuation was unmistakeable. Two days later President Otto Hahn had to tell the Academy of Sciences of the USSR by telegramme that for the time being the Max Planck Gesellschaft could not send any representative to Moscow. Hahn added that he hoped to be able to take up an invitation at a later date.

Hahn set great store on an official Bonn delegation travelling to the Geneva conference, which it would be natural for Heisenberg to lead. Von Brentano, the Foreign Minister, conveyed the Chancellor's wish to Heisenberg. But the atomic physicist did not wish to go. Without giving any reasons, he laconically informed the Foreign Minister that he would neither lead the delegation nor travel to Geneva. As a result the Bonn Foreign Office out of town announced that owing to overwork Professor Heisenberg could unfortunately not represent the Bundes Republic in Geneva.

"That I have not time is clearly nonsense", explained Heisenberg to journalists who ran him to earth in his Ober Bayern home in Urfeld. The atomic researcher who did not speak on the radio and could not go to Moscow gave the true reasons for his refusal. "The advice being given from science is not being sufficiently taken into account by the National Government in the development of the technical means for atomic energy. Science is not being informed about political decisions in this field".

What was behind this statement has only partly become known. It is certain that Adenauer had made decisions on atomic matters over the heads of the experts. So it was, too, with the free for all for the location of the first reactor of the German state. Heisenberg, who had been selected as the leader of the project, had pled for München. Adenauer decided on Karlsruhe. To widespread surprise the Chancellor substantiated his decision with the hint that München lay 'too near' the frontier did not enter into question on strategic grounds. Eventually Otto Hahn weakened to the pressure of the Bonn politician, who had very much wanted to avoid naming another atomic researcher, and took over the leadership of the delegation to Geneva. They set out together, three scientists and two officials of the Foreign Office.

At the First International UNO Conference for Peaceful Uses of Atomic Energy, which was held from the 8–20 August 1955, there were seventy three nations with 1,400 delegates, as well as 4,000 reporters and correspondents. In addition, there came an army of newspaper, radio, and television journalists. Over 1,000 scientific works had been submitted and printed, the greater part of which were presented. Niels Bohr gave the scientific opening lecture. When he went to the lectern to speak on 'Natural Science and the Attitude of Men', he got the assembly to rise in honour and greet the Nestor of Atomic Physics with rousing applause.

Amongst the wealth of scientific news items there did not fail to appear a few surprises during this first great exchange of views. In a meeting of nuclear physicists Soviet scientists published a long list of measurement data. Down to the last decimal, the values agreed with the likewise hitherto unpublished data of the Americans. Was not that a basis for working together in research, it was said in appreciation.

Understandably enough, Otto Hahn often stood at the centre of discussions and in cross-examinations by the press. Even during his first walk, on 9 August, around the atomic exhibitions of the countries participating, he was surrounded by reporters. Many of them had been surprised, for Hahn commented on so few scientific questions, but on so many more political ones. So it was that in a press conference he explained, amongst other things, that no peace will be able to be built out of fear. The exchange of scientific discoveries in atomic research was still a cause of hope because it contributed to reducing mistrust and to bringing an end to the cold war.

All participants had to cope with an extensive programme. Senior citizen Hahn was not excluded. On 11 August he himself led the numerous sessions on the theme 'Processes of Nuclear Fission'. A lively discussion developed. Even Hahn as Chairman had to confess that the rapid advance of nuclear technology was overwhelming for him also. *If I think of with what extraordinarily simple means Fritz Strassmann and I were able in 1938 to prove that uranium had fissioned, it impresses me so much the more to see and to hear in Geneva with what astonishing industrial and technical expenditure and with what cleverly devised and complicated apparata the practical exploitation of the powers slumbering in the atomic nucleus is taken forward today.*

The swimming pool reactor which the USA had put on show in Geneva and which was based on the principle of the fission of enriched uranium was one progress of which Hahn stood in admiration. When the reactor began to operate the bottom of the water pool shone with an unearthly blue light. This effect was based on the so called Cherenkov radiation, named after the Soviet Nobel Prize winner.

At that time one episode appeared in all the newspapers. In front of the swimming pool reactor there stood, as always, a deep circle of inquisitive people. On one day an elderly man of medium height, with bushy eyebrows, and a cigar between his fingers belching smoke, squeezed through the curious crowd. *Can I have a play with your beautiful reactor?*, he asked the operator in perfect English, who busied himself at the control panel behind a glass wall.

"Okay", said the latter. "But do you understand anything about nuclear fission?"

I rather believe, answered the old man, suppressing a smile, *that I did, indeed, discover it . . .*

Hahn was allowed to switch on the reactor. And the American eventually said, embarrassed, "Sorry, but you can't know all the people who are in the atomic business nowadays".

Karl Winnacker's memories of it are revealing. "With his humour and his sound humanity Hahn quickly gained ground at the Geneva Conference to which we other members of the German delegation became very indebted. We even went to the official Soviet reception, at which we were also able to bask in Hahn's fame. This visit took place against resistance from the representatives of the Foreign Office, for the Bundes Republic had no diplomatic relations with Moscow".

At the International Conference for the Peaceful Uses of Atomic Energy held in Geneva atomic researchers from all over the world assembled, summarised Otto Hahn. *Research results which had previously been stamped 'Top Secret' had been received, and were communicated clearly to the public. The Iron Curtain between West and East was lifted for this discussion.*

18
The Call of the Göttingen Eighteen

18.1   Reduced to Silence, Yet not Convinced

At the beginning of May 1955 the Paris Treaty came into force. The Bundes Government had accepted its ratification. Adenauer could now get busy with his long planned changes. The present office of Theodor Blank would be reorganised into the office of the Ministry of Defence. As the new Foreign Minister Adenauer named Heinrich von Brentano. Finally, the last 'Minister without Portfolio', the CSU politician Franz Josef Strauss, also received a department. In November 1954 the Chancellor had already conveyed to him that he was one day to look after 'atomic affairs'... .

In October 1955 Strauss advanced to Chief of the new Bundes Ministry for Atomic Affairs. One of his first official acts was the foundation of a 'German Atomic Commission'[1] of which he named himself Chairman. Officially thought of as the consultation organ for the Minister, this Commission naturally was also a representative of the interests of the industrial and financial worlds which were getting into the atom business with high expectations. Amongst the twenty five members of the Atomic Commission they had the clear majority. Strauss wanted to manage with as few as he could—a few prominent scientists like Otto Hahn and Werner Heisenberg, and two trade union representatives. "I find that is a somewhat under-developed fig leaf", as an SPD member of parliament criticised the composition. Otto Hahn was also elected to be one of the managing

---

[1] *'Deutsche Atomkommission'*.

directors, along with Karl Winnacker, a member of the board of directors of Farbwerke Hoechst AG, and Leo Brandt, the Minister of State. Of this fairly unexpected nomination Otto Hahn thought that at his seventy seven years he was rather too old for this post.

"Here in the House we have different ideas about old age", interjected Strauss, alluding to the eighty year old head of the Bonn government.

"I bow only to your persistence Herr Minister", Hahn said at last.

"Here in the House we also have different ideas about persistence", Strauss informed him once again.

In July 1956 the Nuclear Reactor Construction and Operating Gesellschaft was established, a forty million mark project financed by the state and industry, which should close the atomic 'technological gap' in the BRD. Strauss, who had been voted in as chairman of the supervisory board, voiced at the founding assembly in Karlsruhe his satisfaction at this 'prelude to the atomic age in the Bundesrepublik' which had just taken place. To make the support of commerce and industry more sure, he sped up the German state atomic programme. "It worries me"—and Werner Heisenberg arrived at this assessment at the same time—"that for the people who had made important decisions here, the frontiers between peaceful atomic technology and atomic weapons technology have moved on".

Bonn's first Minister for Atomic Affairs saw his brief not only to be the speeding up of the construction of their own nuclear reactor industry, he pled also for the arming of the Bundeswehr with atomic weapons. Of course, he exceeded his authority and it was not rarely that he was in conflict with Theodor Blank, the Minister of Defence, who appeared more temperate than Strauss. Adenauer finally intervened in the Strauss–Blank controversy himself and appointed Franz Josef Strauss to the post of Defence Minister as a strong man, on 31 October 1956. The new Minister for Atomic Affairs was Professor Balke.

The ministerial change in Bonn was alarming to many at the time, for it was feared that now the atomic arming of the Bundeswehr would no longer be held back. The West German atomic researchers, with Otto Hahn at their head, made a stand against such plans with all their strength on the grounds of political common sense and moral commitment–not least because they feared that they might be drawn into assisting with the military problems of the use of atomic energy. But that they would never ever do—as Hahn and Heisenberg, Weizsäcker, and Gerlach had sworn.

Speaking to a Bonn representative of the GDR press agency[2], Otto Hahn emphatically reinforced this decision. *I believe that the majority of atomic physicists, in any case the well known atomic physicists in West Germany,*

---

[2]The *Allgemeiner Deutscher Nachrichtendienst*, the press agency of the German Democratic Republic (East Germany at that time).

would decline to participate in research work for the production of atomic weapons.

What was to be done in this situation? In a letter of 19 November 1956 the scientists asked the Minister of Defence "to explain publicly that the Bundesrepublik had no intention of either making or storing atomic weapons". Should this warning not be taken seriously, they would say so publicly. Twelve prominent West German atomic researchers, members of the Atomic Commission embodying the circle of those working on nuclear physics, signed this letter. Balke received a copy.

After two months' silence a reply was finally received from Bonn. Strauss, the Minister, invited Professors Hahn, Heisenberg, and von Weizsäcker, as well as Heinz Maier-Leibniz and Wilhelm Walcher, to Bonn to an internal discussion on 29 January 1957.

Interesting details of this 'audience' with Strauss have become well known, in spite of the confidentiality agreed. From the Minister's point of view an excited debate had taken place. In biting terms, Strauss rejected the request of the atomic researchers. As a member country of NATO the Bundesrepublik needed atomic weapons, otherwise they could not negotiate with the Soviets from a position of strength. The rhetoric and manner of the Minister left the scientists with a discordant impression. Hahn and von Weizsäcker hinted that their letter might be published. "If you do that, meine Herren", said Strauss dismissively, "you will be heros from East Berlin to Peking!"

On one February day in 1957 the Lecture Hall of Göttingen University was filled to overflowing. Students from all faculties and members of the teaching staff sat crowded together on the benches or squatted on rough and readily arranged seats. Even so, there were not enough to allow in all those who were interested. A notice on the door of the Lecture Hall announced who had drawn this unusual amount of attention:

Prof. Dr. C.F. von Weizsäcker
'Atomic Energy and the Atomic Era',
II. Lecture: 'Political Effect'.

During the lecture the speaker was interrupted many times by applause, especially when he said, "If small nations, such as Germany, begin wishing to arm themselves atomically, it is criminal and suicidal nonsense".

As far as he knew, so von Weizsäcker continued, no case was known up to now of West German scientists being encouraged to develop atomic weapons. "Nevertheless, naturally we must pose the question of how we should react in the event of such a request. The individual is overtaxed at heart by such decisions. But he has to make it.... I know today that I am not prepared to make bombs. And I also know that others agree with me, such as my distinguished colleague Otto Hahn".

Von Weizsäcker turned to the discussion with the Bundes Minister of Defence. "You will understand if I can not say too much about it here", he explained hesitantly. "But there is one thing I can not keep quiet about.

To our concern we found out that Strauss, the Minister, considers atomic arming to be necessary. If I were to characterise our mood after this conversation I would have to say that we left the Minister silenced but not convinced!"

18.2   A NATO Officer Blabs out of School[3]

On 15 March 1957 the citizens of Germany were taken aback by a report which put all other events in the shade, 'atom bombs are in the Bundesrepublik!'

One of the taboos cherished by Bonn politicians had been broken. The guilty party was the British Air Marshal Percy Bernard[4], Earl Bandon, Commander of a NATO wing stationed in the BRD. In the Headquarters in Mönchengladbach this officer gave away an official secret to representatives of the Dutch press, on 13 March, when he said, "The Fourth Allied Tactical Air Force in south Germany has for some time been in possession of atom bombs and aeroplanes which can deliver them".

From that day the military career of the Earl Bandon was over[5]. Shortly thereafter he was relieved and, so it was said, sent to the Far East. His rash remark, however, set an avalanche in motion. In Bonn and amongst the NATO staff there was widespread annoyance. They sought to trivialise the report: "It is merely a matter of 'tactical atomic shells' and on no account atom bombs". As a denial, it was not enough. Quite the contrary.

So the Bundes Foreign Minister, von Brentano, compared the re-equipping of the army with atomic weapons with the change from the muzzle loader to the breech loader, as it is necessary only once 'in the chain of technical development'. The Chancellor outdid everyone. At a press conference in Bonn on 5 April he made clear the "difference between tactical and strategic weapons" in order to convince the doubters of the harmlessness of each atomic shell:

"Tactical atomic weapons are at root nothing more than a further development in artillery, and it is quite obvious that we can not do without them", said the Chief of the Bonn government with dignity. "These are to be differentiated from the large atomic weapons, which we do not have".

Adenauer's statement put the atomic researchers into a mood of alarm! Almost without pause, Hahn, von Weizsäcker, Heisenberg, Born, and Gerlach telephoned each other. All were agreed that the Chancellor's speech prompted them to active opposition. Hahn requested von Weizsäcker to prepare the text for the long planned action.

---

[3]See footnotes 4, 5.
[4]See Note 12 in Translator's Notes on the Text.
[5]See Note 13 in Translator's Notes on the Text.

On 11 April Hahn waited a long time for a reply in vain. Only towards midnight did the telephone ring. Von Weizsäcker informed Otto Hahn that all the physicists in question, to whom he had spoken at a meeting in Bad Neuheim, were in agreement with the publication of the declaration. A visibly relieved Otto Hahn put the telephone receiver down. Now the action could begin.

18.3 The Göttingen Statement

The text of the Göttingen Manifesto—under which name the declaration of eighteen West German atomic researchers made history—was in the early morning of 12 April 1957 conveyed by telephone to the *Frankfurter Allgemeine Zeitung* and the German Press Agency by a secretary of Otto Hahn. When the employees of the German Press Agency read the protest they were baffled. One hour later they rang the Max Planck Gesellschaft in Göttingen to find out whether the wording was correct. Hahn was able to reassure the caller that the text was right.

By the next day the major newspapers had given the call prime position. The moral dilemma and the sense of responsibility of the solemn appeal of the eighteen atomic researchers had the following wording:

> The plans for atomically arming the Bundeswehr fill the undersigned atomic researchers with deep concern. Some of them informed the appropriate Bundes Minister of their reservations several months ago. Today the debate about this question has become an open one. The undersigned feel themselves obliged publicly to point out a few facts which all those in the subject area know, which, however, do not appear to be sufficiently well known by the public.
>
> 1. Tactical atomic weapons have the destructive effect of ordinary atomic bombs.
>
>   By 'tactical' one means, to put it briefly, that it is used not only against housing but also against troops in ground battles. Each individual tactical atom bomb or atomic shell has a similar effect to that of the first atom bomb which destroyed Hiroshima. Since tactical atomic weapons would nowadays be available in greater numbers, their destructive effect would be very much greater. By 'small' it is meant that these bombs are small in effect only in comparison with that in between the 'strategic' bombs developed, especially the hydrogen bomb.
>
> 2. There is no known natural limit to the ability that can be developed of strategic atomic weapons to wipe out life.

Nowadays a tactical atomic bomb can destroy a small town; a hydrogen bomb, however, would make an area the size of the Ruhr region unliveable for a time. By the spreading of radioactivity the population of the Bundesrepublik could probably be wiped out today. We know of no technical means of protecting the greater part of the population from this danger.

We know how difficult it is to draw political consequences from these facts. It is wanted to deny this right to us as non-politicians; our job, which is the activity of pure science and its applications, and in which we lead many young people into our field, does, however, put upon us a responsibilty for the possible consequences of these facts. We can not keep silent about all political questions. We profess our faith in our freedom as found in the western world in contrast to communism. We do not deny that the mutual fear of the hydrogen bomb has brought about today a considerable contribution to the maintenance of peace in the world and of the freedom of a part of the world. However, we hold that this way of securing peace is in the long run inadmissible, and we hold that the danger, if it fails, is fatal.

We feel no competence to make concrete proposals to the politics of the great powers. For we believe that a small country such as the Bundesrepublik is today best protected and world peace best promoted if it expressly and willingly forgoes the possession of atomic weapons of any kind. At any rate, none of the undersigned is prepared to take part in the production, the testing, or the deployment of atomic weapons in any way.

At the same time we emphasise that it is exceedingly important to promote the peaceful applications of atomic energy with all means, and we desire to help in this task as we have done hitherto.

Fritz Bopp	Heinz Maier-Leibnitz
Max Born	Josef Mattauch
Rudolf Fleischmann	Friedrich Adolf Paneth
Walther Gerlach	Wolfgang Paul
Otto Hahn	Wolfgang Riezler
Otto Haxel	Fritz Strassmann
Werner Heisenberg	Wilhelm Walcher
Hans Kopfermann	Carl Friedrich Frhr. von Weizsäcker
Max von Laue	Karl Wirtz

18.4 'Unsuspecting Fools'

It is to the credit of the Göttingen eighteen to have explained to the German people the possible terrible consequences of keeping atomic weapons in their country. The most lasting effect, of all, of their statement was that tactical atomic weapons were equated with conventional atomic bombs.

The scientists solemnly reaffirmed that they would take no part in the production, testing, or deployment of atomic weapons. That lent a great moral weight to their appeal.

Certainly, many a version of the manifesto appeared as a compromise formula. Understandably, note was taken of the party political interests of the eighteen professors. The majority of the signatories stood nearer to the CDU rather than the Social Democrats or the Free Democrats. Strassmann was one of the first to take the critics to task; versions stating that the physicists did not want to take part in politics did not find his approval. "Why should I as a scientist not have a clear political view?", maintained Strassmann. "I have the ability to think politically. It forbids the use as well as the nuclear arming and deployment of nuclear weapons, especially in a country like the Bundesrepublik".

Especially did the politically explosive phrases of the manifesto—a small country like the Bundesrepublik must emphatically and willingly give up the possession of atomic weapons of any kind—agitate the leading Bonn politicians.

At midday of 12 April, a few hours after the telephone delivery of the text, Strauss had already learnt of the action of the eighteen disobedient professors. Immediately he rang Otto Hahn and heaped reproaches upon him. The scientist recorded this conversation in a note of keywords. *12*th *April, 1957. In the early afternoon Strauss phoned me. For more than an hour he was very angry and carried on. Uncouth man! We had broken a gentleman's agreement, etc.. I contradicted, so it's going well...*

Strauss did not take Hahn's practical hint. He repeated much more to the press; the action of the professors had been "an ill considered experiment", for it had been sent "to the wrong address", to the government of the Bundesrepublik, which had solemnly foresworn the production of atomic weapons... . Hahn rejected this accusation publicly. *I can send it only to the proper address, which I have done.*

In an interview with the magazine *Der Spiegel* Strauss characterised the atomic researchers as "unsuspecting fools". He spoke disparagingly in the Bonn Press Club about the President of the Max Planck Gesellschaft. Hahn was "an old wally who can not hold back his tears or sleep at night if he thinks of Hiroshima".

On 12 April Adenauer was in Eichholz near Bonn when the news of the appeal of the Göttingen professors took him by surprise. A statement was expected from the Head of the Government, which was obviously not easy for him to do. Journalists present noticed that the eighty one year old Chancellor was "as white as chalk", had "unusual signs of great excitement", and spoke "with a trembling voice". "If the scientists say", said Adenauer making his displeasure known, "a small country like the Bundesrepublik best protects itself by willingly abandoning atomic weapons, then I say that such a declaration has nothing to do with scientific knowledge. It has the nature of foreign policy. To make their judgment knowledge must be had which these gentlemen do not possess. For they have not come to me".

Such a presumptuous stance brought forth a storm of indignation within and without the country. The Paris newspaper Le Monde called Adenauer's statement "a masterpiece of ineptitude and arrogant insolence". Erich Ollenhauer, Chairman of the SPD, saw in the Chancellor's words "an unsurpassed measure of arrogance and superciliousness". Even Balke, the Bundes Minister for Atomic Affairs, distanced himself—for many of his loyal party supporters had been surprised—from such unqualified comments and welcomed the Göttingen Manifesto.

Amongst the eighteen West German academics who had signed the Göttingen Appeal one missed the names of a few well known atomic researchers such as Diebner, Harteck, and Jordan. They did not sign; better to say, they had been deliberately not asked for their signatures. Bagge might perhaps have been a nineteenth signatory to the appeal, as he confirmed, but at the time he was abroad.

In his paper 'The Failed Revolt'[6] of 1956 Jordan had stated that "a hot atomic war would be limited to about thirty six hours" and then the time would come "to look at the costs". He considered that "mankind" might be "reduced to a few thousandths of its original numbers". But what does it matter? A thousand years later the "problem of over-population" would again be current. Probably then "humanity would have long since prepared to stay underground again for five years until the atomic stench outside had died away". Besides, it is well known from history that once before— in the twelfth century—man had been warned of the lethal effects of a new weapon, namely, the cross-bow. Even so, mankind came to understand the cross-bow.... Jordan, who was trying to get a seat in the Bundestag, did not shrink from attacking his scientific colleagues and from "political unworldliness".

After the Göttingen Appeal became public knowledge Otto Hahn received countless statements of solidarity from all around the world. There were days on which Hahn and his secretary were busy only with the sorting of the post. The Manifesto of the Göttingen eighteen had found an exceptional response, and abroad also.

Amongst the letters recognised which Hahn received was a letter from Joliot-Curie. "On the other hand we know that we can not remain indiffer-

---

[6] 'Der gescheiterte Aufstand'.

ent to what has happened to us. Permit me to say how very much we here greet the appeal which you with your seventeen colleagues have signed".

Responsible scientists all over the world felt strengthened through the action of the Göttingen eighteen. The Japanese Nobel Prize winner for Physics, Hideki Yukawa, sent a resolution to Hahn in May, 1957. The statement of the Göttingen scientists has had an "extremely encouraging" effect upon Japanese atomic physicists, who have pursued this same goal in their own country.

Albert Schweitzer, the world famous humanitarian[7], gave a strong warning on 23 April of the dangers of the bomb tests carried out. Countless radio stations around the world transmitted his talk. Otto Hahn vigorously welcomed this call, for it was a warning out of a mouth qualified to speak.

18.5 Ten Against Five

Göttingen academics had already caused a sensation in 1837 by their revolt against the sovereign prince[8] of their land. Seven professors, amongst them the brothers Grimm, at that time protested against the arbitrary exercise of his power by the new King of Hannover, who had refused to respect the constitution. The monarch immediately relieved the rebelling professors of their positions, and three of them, Friedrich C. Dahlmann, Jakob Grimm, and Georg G. Gervinus, had to leave the kindgom within three days.

"It did not befit the subject with his narrow-minded view to lay down the standards in the business of the head of state, nor to arrogate to himself a public judgement of legitimacy of the confinement of his high spirits in a darkened cell". With these words, the Minister of the Interior, Gustav A. von Rochow enacted the dismissal of the 'Göttingen Seven'. What would the fate of the 'Göttingen Eighteen' be? The Bundeskanzler had reprimanded his eighteen recalcitrant 'subjects' because of their 'narrow-minded view' in political matters.

On 15 April four of the eighteen professors, namely Hahn, Heisenberg, von Laue, and von Weizsäcker, received by telephone from the office of the Bundes Chancellor an invitation to a discussion. In his position Otto Hahn could not decline. He proposed as reinforcement Professor Gerlach as a further partner in the discussion. Heisenberg refused to appear in the audience with the Bundes Chancellor and announced himself to be "ill". When Adenauer heard of it he rang up the atomic researcher but could not change his mind. In the place of Heisenberg, Professor Riezler was finally invited to Bonn.

---

[7] See Note 14 in Translator's Notes on the Text.
[8] 'Landesherr'.

Understandably, the summons of the professors to the Chancellor aroused a considerable stir. Did Adenauer want to see to a retraction? Otto Hahn and his colleagues said, however, that nothing that had been said by them would be taken back. They were going to remain firm.

In the Schaumburg Palace, the official residence of the Chancellor, the coming and going of diplomatic visitors and the holding of receptions was the usual picture. But what came to pass in the morning of 17 April 1957, doubtless surpassed the measure of routine diplomatic work. The Bonn government's palace resembled a besieged fortress. Reporters and photographers from the press, radio, and television pressed in great numbers before the entrance and hindered the continuous coming and going. Luftwaffe officers of the armed forces hauled large ordnance survey maps.

At last, the first diplomatic car drove up. The Minister of Defence, Strauss, and Generals Adolf Heusinger and Hans Speidel stepped out, and went briskly into the Chancellor's palace. But the gentlemen of the press had not been waiting for them. Their interest much more was in the two Mercedes which shortly drove up, a few minutes before ten o'clock. The two cars were immediately surrounded so that those inside could hardly get out and only with difficulty could make their way to the entrance.

Otto Hahn, the senior one of the groups, had their undivided attention. *Only take plenty of pictures of us*, were his words, and with a typical glance over the Government Palace he went in, *Who knows whether they will let us out again*.... Thereupon Hahn disappeared with his companions, Professors Gerlach, von Laue, von Weizsäcker, and Riezler, into the Bundes Chancellor's palace.

Patiently the press people waited for the reappearance of the scientists. They busily debated the chances of the five and prepared for the detailed press discussions. The hours went by and nothing happened. At last the waiting reporters—for sixteen hours had gone by—were informed that the Chancellor had imposed a news blackout. A limited range of questions would be answered by the Bundes Press Chief. It would not be possible to see or speak again to the scientists. In addition they had declined to be photographed together with the Chancellor.

The press did not know what to make of these words. At some time, the academics had to reappear. The Chancellor's residence was not a middle ages stronghold behind the wall of which the lord of the castle could make his displeasing guests disappear. Since those days, the times had completely changed....

In fact, the reporters were not to see the scientists again that day. The Chancellor had let out his guests through a side door after a seven hour long conference. Prince Bernhard of the Netherlands, who had thought to pay a courtesy visit to the Bundes Chancellor, was the only one whom the journalists set eyes on. He also had to wait until the Chancellor had time to see him. That the impressive press contingent was not waiting on his account, as the Prince first assumed, he must have discovered to his great regret.

The five professors had to face at the oval conference table a superior strength of ten 'opposite numbers'. Opposite them sat Adenauer, Strauss, the Permanent Secretaries Hans Globke, Walter Hallstein, and Josef Rust, the Chancellor's official Franz Haenlein, the Chancellor's personal adviser, Hans Kilb, the Press Chief Felix von Eckhardt, and Generals Heusinger and Speidel.

It must have been depressing that the only advising participants, such as Globke and the two generals, had held leading positions in the time of the Nazis. As Lieutenant General and Chief of the Operations Unit, Heusinger had been responsible for Hitler's deployment plans. For the time being he exercised the function of an inspector general of the Bundeswehr[9].

Speidel, Chief of Staff of an army group in the Second World War, had recently been appointed as Commander in Chief of the NATO Land Forces in central Europe. Not two months after the meeting with Adenauer, the two, Speidel and Heusinger, were promoted to four star generals, the highest officer rank in the Bundeswehr.

Globke had covered himself in guilt through his complicity in the persecution and eradication of the Jews. Kilb was later to cause a stir when he was charged with accepting large bribes and had had to be taken into custody.

The Chancellor opened the meeting. To the scientists his three quarters of an hour long speech about the 'threatening communist danger' and the 'need of strength in NATO' brought nothing new. At best, it enlarged the vocabulary of the cold war with further nuances. Adenauer's conclusion was that the Bundeswehr needed atomic weapons. And with mock surprise the Chancellor said that he really could not understand the scientists, for the entire disaster with the atom bomb had been brought about by one of their number, Einstein.

The subsequently short appearance of the military, who soon after had flown back to NATO Headquarters in Paris, had turned into a spectacle for them. With the help of large sized maps of Europe and the world, Heusinger had sought to make it clear why the Bundesrepublik had to have atomic weapons, because of the prevailing military balance of power. Opposition to nuclear weapons was nonsensical, the General judged. Also, in the First World War there had been reservations about the use of machine guns. Since then it had become a standard military armament.

Speidel insisted on the principle of 'deterrence' with atom and hydrogen bombs. "The feeling of not having the best weapons kills the spirit of a soldier".

---

[9] *Translator's Note:* The Armed Forces of West Germany.

Strauss was agitated, and in little control of himself. "Never", he exclaimed, "will a renunciation of atomic weapons be pronounced by the Bundes Government".

Up to then the five scientists had followed the proceedings in silence. "Now we wish", Adenauer, to whom the one-sidedness of the negotiation proceedings had begun to stand out, at last said, "to hear the professors". Hahn spoke first. He summarised the standpoint of the scientists which had been set out in the Göttingen Manifesto. The signatories to the appeal, Hahn said, would not move from their opinion. Von Weizsäcker joined in and resolutely rejected the accusation of disloyalty levied by Strauss. The publication of the appeal had been provoked by Mr. Chancellor's statement about tactical nuclear weapons. "When we read this", said von Weizsäcker fearlessly, "we went up the wall".

Adenauer's face looked like a mask.

18.6   An Incorrect Communiqué

The Chancellor insisted on a joint communiqué. But only after many changes, deletions, and additions did the professors accept a draft which von Eckhardt, the Press Chief, had prepared.

"The object of the discussion was the world political and strategic situation in the atomic era in connection with the statement by the eighteen German scientists of 12 April 1957", it said on the paper. To general surprise, it was asserted that the Bundes Government "shares the concerns which are expressed in the named statement; it agrees with the motives and goals of the scientists". That the Bundes Government makes no atomic weapons and will not approach the atomic scientists for doing so was verbosely assured, likewise.

In contrast, the atomic arming of the Bundeswehr was not mentioned. Adenauer and Strauss would not get involved in that at all, for in this matter they could not agree with the scientists' opinion and the Göttingen statement. The changes in the communiqué demanded by the scientists were obtained only against the resistance of the Minister of Defence.

There was, last of all, an amusing incident in the reading of the newly written version. Adenauer, who did this himself, came to a halt at one point. "A comma belongs here". Hallstein contradicted him and referred to the rules of grammar. But the Chancellor remained obstinate, "But a comma *does* belong here!"

Somebody made a comment, in a low voice so that all heard, *Where the comma is, I decide!* With an evil look, Adenauer sought to find the utterer. He did not succeed. But Gerlach laughed to himself. He had recognised straight away the Frankfurt accent of Otto Hahn.

At the end of the discussion with Adenauer Otto Hahn felt, retrospectively, there had been a positive outcome. *We are content with the results—*

although Minister Strauss less so—for we could not achieve any more, and we held our position. Doubtless there is the ring of a certain resignation to these words. Otto Hahn and his colleagues came to know the limits of their influence on policy.

From this assessment it will not be able to be forgotten that the scientists were certainly relieved when they were saved from personal reprisals. Professor Hahn frankly admitted afterwards, when the talk with the Chancellor suddenly arose, *I was really worried.*

19
Against Nuclear Weapon Experiments and Nuclear Balance

19.1   Political Creeds

Thanks to his moral integrity Otto Hahn was trusted everywhere. He used it to point uncompromisingly to three important goals. For him the cessation of nuclear weapon tests, not transferring atomic weapons in order not to let the number of atomic powers become larger, and general disarmament were the essential challenges. Hahn occasionally emphasised that he was not a politician. Yet his speeches, appeals, warnings, and his appearance in public betrayed a purposeful political engagement. His distinctively humanitarian convictions directed him logically to this path.

The decorations and honours which Otto Hahn had received during his life made a long list. He especially appreciated the conferral of an honorary doctorate from the famous University of Cambridge, that last place of work of his teacher Rutherford. When Hahn travelled to Cambridge in June 1957 to receive the honour it was at a time when the hydrogen bomb test of the new atomic power Great Britain was to be carried out near Christmas Island. During his visit Hahn joined the crowd of protesters, and also demanded that this test be called off.

For three days Otto Hahn was at the centre of public interest in London and Cambridge. The English press reported columns about his life and work. "Otto Hahn is one of the figures of world history", declared *The Observer* to its readers on 9 June 1957, describing his life. "But he possesses none of the attributes of the traditional luminaries in history books. His slight, somewhat bowed figure, which with its high brow has the effect of his

features having been carved, with his expression of searching honesty and
critical inviolability, have something of a boundless nobility about them".

13 November 1957. The Grand Concert Hall in Vienna was not big
enough to hold everybody interested who wanted to hear Professor Hahn's
lecture 'Atomic Energy for Peace or for War?' The speaker received a storm
of applause. He was was received both as an outstanding scientist and the
tireless admonisher of atomic war and atomic death. Visibly moved, Hahn
turned to his audience. He called upon them to join with him to bring
about a conclusive end to all atom and hydrogen bomb experiments. The
radioactive dust of death had already gradually contaminated the entire
Earth's sphere, bringing men illness, hereditary defects, and sterility, and
could finally lead to their elimination without there ever having been a
'hot' atomic war. In his opinion an international congress of scientists must
prepare for the first step in reaching an agreement on the cessation of tests.
May the recognition grow, said Otto Hahn, closing his rousing lecture, *that
with the possibility of destroying all life on earth which exists today, a large
scale "war is no more the continuation of politics by other means"*.

Once again in Göttingen, Otto Hahn found amongst his voluminous post
a letter from Pasadena, California, dated 6 November 1957. Linus Pauling
was the writer. He informed his German colleague of a promising action.
Within a few days Pauling had succeeded in obtaining the signatures of
over 2,000 American scientists to an appeal which demanded an agreement
for the immediate end of the use of nuclear weapon tests. Prompted by this
unusual response, Pauling intended also to win over scientists abroad.

"Lieber Herr Professor: Ich hoffe, dass Sie unterzeichnen werden!" [1] Pauling's postscript to the letter, written in his own hand, showed how very
helpful to him the involvement of his German colleague would be. Hahn
did not hesitate for one moment. *I will sign* he noted spontaneously in the
letter's margin.

In all, 9,235 scientists from forty four nations put their signature to the
Pauling Appeal, which the initiator presented to the United Nations on 13
January. Below it were also found the signatures of well known scientists
from the USSR, such as the President of the USSR Academy of Sciences,
Alexander N. Nesmeyanov, the first Soviet Nobel Prize winner for Chem-
istry, Nikolai N. Semeonov, and Rutherford's pupil, Peter L. Kapitza.

To this end, in July 1957 the first of the later traditional Pugwash Conferences was held, at which scientists from the West and East gathered to
discuss together the cessation of atom bomb tests and questions of disarmament.

Hahn powerfully outlined his viewpoint on the question of the non-
proliferation of atomic weapons in an interview which he gave to the Danish
newspaper *Politiken* at the beginning of December 1957. The peace of the

---

[1] *'Dear Professor, I do hope that you will sign!'*.

world also hung, said Hahn, upon the possession of atomic weapons remaining limited to the great powers. A *little Hitler* somewhere in the world could plunge the world into destruction.

19.2   Honours and Prizes

The year 1958 brought Otto Hahn a reunion with East German friends and colleagues, with Gustav Hertz, Max Volmer, Manfred von Ardenne, and Max Steenbeck. The cause of the celebration was the one hundredth birthday of Max Planck, at the invitation of the German Academy of Sciences in East Berlin. As President of a scientific society which bore Planck's name, Hahn was pleased to take up the invitation. He alone was a member of the Academy, although it had been suggested to him by the western side that he give up this membership. *I shall remain in it, because I believe it right to do so*, was Hahn's short reply, putting up with no comment to the contrary.

To those celebrations, which took place on the April 24 and 25 and the high point of which was a ceremony in the State Opera House Unter den Linden, Lise Meitner travelled from Stockholm. From the West German side, in addition to Otto Hahn, Max von Laue and Max Born also took part. Hahn's message of greetings was received with great applause when amongst other things he explained, *The spirit embodied in Max Planck of reverence for truth and his uncompromisingly maintained humanity will find its echo in all scientists today.*

At the get-together Otto Hahn, Gustav Hertz, and Lise Meitner exchanged memories of their work together at the University of Berlin. They talked about the wooden workshop and mesothorium, and agreed that the achievements of those times now only counted as the history of radium research.

Hahn and Meitner later seemed pleased when they found out that these memories had been nourished, for some years later, on 15 November 1966, at the Institute of Chemistry of the East Berlin Humboldt University, in Hessische Strasse, a commemorative plaque revealed:

> In the former wooden workshop on the ground floor of this building the radium researchers Otto Hahn and Lise Meitner made important discoveries in the natural sciences from 1906/07 to 1912.

One year after the commemorative ceremony in honour of Max Planck, the German Academy of Sciences reciprocated the visit of the President of the Max Planck Gesellschaft. An anniversary was again the cause, for on 8 March 1959 Otto Hahn celebrated his eightieth birthday. The Academy's President, Werner Hartke, presented to the birthday boy the highest decoration which the Academy was able to give, the Helmholtz Medal. The last time this medal had been awarded was to Röntgen in 1919. That Hahn was especially delighted by this honour is undisputed. From the hands of the Bundes President Heuss, who had travelled from Bonn, he received the highest decoration of the Bundes Republic, the Grand Cross of the Order of Merit[2].

A quite different kind of honour which should have come to him Hahn was just able to avert. On 1 July 1959 the new Bundes President was to be elected. Up to the very last moment it remained open who was to be the successor of Theodor Heuss. To the annoyance of his loyal party members, for a long time Adenauer could not make up his mind whether he should now remain Chancellor or seek the presidency. The FDP did what it could to nominate another candidate, Otto Hahn. But the scientist dashed this intention before it became 'official' with a little joke: *That can not possibly come into question. Two eighty year olds in Bonn? One is more than enough!*

It was with satisfaction that Otto Hahn noted a real decoration: a ship was to be named after him. The 'Otto Hahn' was a freighter of 16,900 tons, 172 metres long, and which for 100,000 miles at sea required about 10 kilogrammes of uranium 235. For the same distance the ship would need some 70,000 tons of coal. Otto Hahn received the launching of this first atomic powered ship of the BRD on 13 June 1964, with delight. In it he saw an example of technical progress in the peaceful application of atomic energy, which he had helped be developed through his work.

19.3   A Departing President

For Otto Hahn's social activities his position in the Max Planck Gesellschaft was not always beneficial. In order not to have to avoid burning topical questions Hahn stressed on many occasions that at that moment he was not speaking as President but as a private person. In this way he sought to gain some freedom of action, for as the President he was bound to the agreements reached by the governing body.

The senate is the most important organ of the Max Planck Gesellschaft because it is the one which makes the decisions. It stands parallel to the Kaiser Wilhelm Gesellschaft if one examines its composition. So we find amongst the forty senators of the year 1958/59 only fourteen scientists. Representatives from the worlds of commerce and industry, banking, the governments of the State and the Lände, together with the current Mayor of West Berlin, represent the majority. The Bundes Government has the financial care of the Gesellschaft combined with the claim of a 'place and voice' in the senate. Despite this, the state subsidy for the Max Planck Ge-

---

[2] *'Das Großkreuz des Verdienstordens'.*

sellschaft is expected, as also later lamented by the Society of its budgetary policy, to foster science from the same budget allotted to it, as for instance happened in the support of the dairy industry.

As President, Otto Hahn was also constantly confronted with financial worries. *Over the next years thousands of millions will be spent on arms. Should it not be possible to find a few hundred million for research, science, and schools?*

Hahn voiced these sentences at the main gathering of the Max Planck Gesellschaft in 1956—or, more correctly, had wanted to voice them, for at the last minute he refrained from them. In the advance material for the press, which published the complete text of the speech, the remark about the thousands of millions for arms is included.

The year 1960 was, in some respects, the last turning point in Hahn's life. It brought him not only his farewell as President of the Max Planck Gesellschaft but also personal sorrow. In April he stood moved at the grave of a dear friend, and said, filled with grief, *With the demise of our dear Max von Laue one of the last of the circle of friends of the year of 1879 is gone from us . . . .* The last of all was Hahn himself. Four months later he was hit by one of the heaviest blows of fate. Hanno, his only son, and the latter's wife, Ilse, died unfortunately in a car accident. It was also with this painful event that Hahn's recollections in his autobiography *Mein Leben* suddenly broke off . . . .

19.4 Last Journeys

With advancing age it was no longer possible, leaving aside a few exceptions, for Otto Hahn to fulfil his countless invitations to lectures and celebrations from all over the world, as he had earlier liked to do and had regularly done.

Together with his son Hanno, an art historian by profession, Otto Hahn spent a few weeks in Israel in December 1959. It was his last great journey. However, it did not take a private, but rather an official, character. As the MPG's President, accompanied by the biochemist and later Nobel Prize winner Feodor Lynen as well as the nuclear physicist Wolfgang Gentner, Hahn took up an invitation to the Weizmann Institute of Science in Rehovoth.

The broader significance of a tour of the nuclear research institutes there and of the Israeli atomic reactor was that this visit fulfilled a function of drawing the peoples together. Otto Hahn in Israel—that was the starting pistol for a German–Israeli collaboration in the scientific field, many years before diplomatic relations were established between the two states, until most recent times of the greatest strains, when they were broken off. Otto Hahn was delighted with the example of his mission to Israel, being able to experience afresh initiating collaboration, showing that science is international, lives on the free exchange of ideas, knowledge, and persons, and helps overcome political differences.

The story of Joachimsthal is instructive. One day the little town in the ore bearing mountains of Bohemia, which had dazzling wealth as the result of a rich silver deposit—the Joachimsthaler was a much sought after means of exchange—sank to insignificance when the supply of ore was exhausted. After the discovery of radioactivity Joachimsthal gave thanks that its rich uranium deposits gave it a new heyday. As a radium spa it has a world wide reputation. As a tribute to the researcher couple Marie and Pierre Curie, who had discovered radium and thereby had contributed to the fame of the town, a memorial and commemorative plaque were to be unveiled in July 1966.

With a little scepticism Otto Hahn travelled on 10 June from Göttingen, for he felt himself a member of a nation which had inflicted suffering on the Czechoslovakian people during the Nazi period. At that time there were no diplomatic relations between the two states. Hahn was therefore pleased at his hospitable reception. *I have not recovered from my surprise at the friendly reception which I have met everywhere*, he recounted to the newspaper *Lidova Demokracie. My opinion is that personal meeting places of this kind are the best route to the removal of all misunderstandings and to the creation of good relations which lead with certainty to an enduring peace.*

As the guest of the Czechoslovakian Academy of Sciences, at the end of his journey Otto Hahn took up an invitation to Prague. In an address over the Czechoslovakian radio Hahn expressed his opinion about the relations between the two countries. According to a contemporary report Hahn showed an "astonishing sense of the realistic and idealistic values which are able to bind the peoples together".

19.5   Atom Bombs Stop

When would the nuclear powers give practical realisation to their preparedness to bring an end to their atom bomb tests. This question repeatedly occupied Otto Hahn in the last years of his life. There seemed to be some progress when for the first time atomic experts from the USA, England, the Soviet Union, France, Poland, Rumania, and the ČSSR met in Geneva and started their work. During the conference, which took place from 1–20 August 1958, they came to the result that nuclear explosions were detectable over the whole world, therefore violations of a test ban would always be detected.

Otto Hahn had already recommended such an international conference of scientists earlier. The discussion in Geneva must have filled him with satisfaction. They would certainly have been under pressure as a result of the incessant nuclear weapon tests.

This menacing attendant circumstance made clear to the world that an arduous path would have to be covered to reach an international agreement on the cessation of nuclear weapon experiments. That was something of which Otto Hahn was also conscious. *Why are there always more tests when so many hydrogen bombs are lying ready, and which are able to snuff out life on the Earth?*, he lamented at the meeting of the Max Planck Gesellschaft in July 1959.

After year long exchanges the USA, the USSR, and Great Britain at last found themselves at the end of an agreement about a stop to testing. At this point almost five hundred nuclear weapons had been exploded, four hundred and thirty of them above ground, and had radioactively contaminated the biosphere of our Earth.

The 'Treaty on the Banning of Nuclear Weapons Testing in the Atmosphere, in Space, and under Water' which the Foreign Ministers of the three states signed in Moscow on 5 August 1963, represented progress. For the first time one had succeeded in curbing one branch of the atomic arms race even if the two atomic powers France and China, who had not signed the agreement, were to continue their early tests. Otto Hahn warmly greeted the cessation of nuclear weapons tests. He hoped for further progress in disarmament, Hahn stressed in a press survey.

Thanks to a world wide protest and diplomatic efforts, an agreement was at last reached on the question of the non-proliferation of atomic weapons. On 11 July 1968 the governments in Moscow, Washington, and London signed the text of an appropriate treaty. On the same day a further three hundred and thirty states signed. The Bundes Republic of Germany, however, hesitated to honour the atomic ban treaty, which Strauss denigrated as a 'Super Versailles'. Owing to political considerations West Germany deposited the document of ratification only in 1975.

Otto Hahn did not witness the conclusion of the atomic weapons non-proliferation treaty, for at that time he had lain seriously ill for months in a Göttingen hospital. A few days later, on 28 July 1968, this pioneer of atomic research died, who to the end had pled for the exclusively peaceful use of atomic energy. When the press agencies around the world reported the passing away of Otto Hahn the loss suffered by humanity was felt in East and West.

"The number of those who had been able to be near Otto Hahn is small", wrote Fritz Strassmann in an obituary. "His behaviour was completely natural to him, but for the next generations he will serve as a model, regardless of whether one admires in the attitude of Otto Hahn his humane and scientific sense of responsibility or his personal courage".

19.6   The Sun on the Earth: The Responsibility of Science

Until shortly before his death Otto Hahn had emphasised that for him the Göttingen Appeal was just as valid as before and had lost nothing in actuality. The other signatories shared this view. In 1961 von Weizsäcker explained that if he had the choice once again to sign or decline, he would without hesitation sign again. And Fritz Strassmann, when he was asked about it in 1967, said that the last ten years had strengthened him in his conviction that this Appeal had been needed.

When von Weizsäcker a little later offered a political compromise and said that one must learn to live with the atom bomb, he plunged Otto Hahn into disbelief. For Hahn *wanted to save from the Statement of the Eighteen what was to be saved*, as he informed the wife of Max Born in August 1962 from his holiday resort of Garmisch. *V. Weizsäcker has actually caused confusion, and I have told him of my regret about his very diplomatic work "Mit der Bombe leben"*[3]... *(but I believe) that he still feels bound to the Statement of the Eighteen.*

Vigilance in the spirit of Otto Hahn's endeavours must similarly be called upon for the disputes which continually flare up over the atomic weapons non-proliferation treaty, and which require infringements to be examined. For instance, South Africa, which exploded its first nuclear bomb in the autumn of 1979, and which was not one of the signatories of the treaty on the non-proliferation of nuclear weapons. Certainly Otto Hahn made a carefully considered decision when in 1964 he refused the invitation of the government of South Africa to officially open that country's first atomic reactor.

"Can one, may one, hold the researcher responsible for the consequences of his work? Everyone who knew Otto Hahn knew with what unsparing clarity he had put this question to himself. We honour him, who as a researcher in his work, just as much as a man in his thought and deed, was ever a model of uprightness and conscientiousness, and even more so by the questions and answers he raised and gave by virtue of his personal conduct". These words were spoken by the atomic physicist Max Steenbeck in 1964. Together with Manfred von Ardenne, Gustav Hertz, and others, Steenbeck had been involved in the construction of the Soviet atomic bomb during their internment following 1945.

The accusation had been made against Otto Hahn that he ought to have kept the discovery of the fission of uranium secret or should have consulted his conscience before making it known. Steenbeck, who had studied the philosophical–ideological question of responsibility in our time, came to

---

[3] *'Living with the Bomb'.*

the only right answer, "Without knowledge an appeal to the conscience is sentimental prattle".

It would have been too simple to want to put the responsibility on the scientists alone. The broadening of our perception and knowledge of the material world in which we live is necessary for the existence of mankind. The search and gaining of knowledge are therefore never able to be damnable. The moral categories of 'good' and 'bad' do not hold for a discovery or invention made at the hand of man. What is crucial is for what purpose society and man in society have chosen to use it. Hahn's coworker Fritz Strassmann voiced this to the author in 1977, "It is not a discovery which is good or evil but rather what men make of it".

For the researcher of today there arise, as Steenbeck set out, two fundamental demands: the obligation of absolute truthfulness; and the duty to be of use to human society.

Otto Hahn himself often came to doubt whether it had been right to have closed the path for mankind to the production of atomic energy. *For years I have occasionally thought over whether it had been better that the entire utilisation of atomic energy had never come into being*, Otto Hahn had said of such considerations in a speech at the annual gathering of the Max Planck Gesellschaft in the summer of 1955. *Then on the other hand one admittedly must wonder whether when the natural fuels of petroleum and coal were being cut back on, it was not just lucky that we succeeded in snatching its secret from the atom.*

Hahn certainly saw in the production of energy by fusion of the light elements into helium a definitely better solution for the future. In an energy releasing fusion reactor there are produced neither dangerous radioactive fission products nor the bomb explosive plutonium. Only then would atomic technology, in Hahn's view, be a blessing for mankind. *Uranium, if no longer needed for war and no longer needed for atomic power stations, would become superfluous, and could be sunk in the sea....* In these words Hahn also gave expression to his secret hope of finally getting rid of *the curse of uranium* by a solution brought about by 'technical' progress. Otto Hahn also said much the same in an interview of 30 May 1958, which appeared in the *New York Times*.

In his sizeable important lecture 'On the History of Uranium Fission and the Consequences Arising from Its Discovery'[4]—held on the 100th gathering of the Society of German Natural Scientists and Medical Doctors[5] on 30 September 1958 in Wiesbaden—Hahn used just as insistent words. *Today we have the hydrogen bomb as the threatening spectre of the explosive union of hydrogen and helium. But our Sun shows us something rather*

---

[4] *'Zur Geschichte der Uranspaltung und den aus dieser Entwicklung entspringenden Konsequenzen'*.

[5] *'Gesellschaft Deutscher Naturforscher und Ärzte'*.

different; we owe to its thousands of millions of years of regular fusion of hydrogen into helium that our Earth is inhabitable, and that it did not chill to a heap of stones long ago.

Hahn's faith that in the future thermonuclear energy would be tamed remains unbroken to the end, despite all critical assessment to the contrary. *Our children or our grandchildren will master the process; they will bring the Sun to the Earth—if they are permitted further to live on the planet.*

We are forced to respect his far-sightedness in speaking such words. It is based on the realisation of how dangerous can be plutonium produced in nuclear power stations if it falls into the hands of terrorists, and how dangerous the radioactive waste can be. Today humanity sees how threatened it is not only by the nuclear arms race but by the atomic over-kill potentials of the great powers, especially the increasingly seen risks growing with nuclear power stations. The euphoria of the previous years over the civilian use of the taming of atomic energy has evaporated, especially since the catastrophic reactor accident at Chernobyl in the Ukraine in 1986.

To a not unforeseeable extent the peaceful use of atomic energy has led to discord. Reactor safety and radioactive waste disposal have not been overcome to the present day. The most fateful discovery in the history of the natural sciences has presented mankind not only with the portent of atom bombs of Hiroshima and Nagasaki, *uranium is a curse... .*

As we must conclude, Otto Hahn is not to be held personally responsible for the consequences of his discovery, but he suffered from them and because of the constantly smouldering conflicts of conscience became a tireless watchman for the world of a life worth living, at peace, without anxiety caused by the atom. His engagement with science, humanity, and peace is also exemplary and to be remembered for following generations.

These sentiments were reflected in the establishing of the Otto Hahn Peace Medal on the occasion of the 50th anniversary of the discovery of uranium fission. This decoration, a medal of solid gold, has since then been awarded by the German United Nations Society[6] "for outstanding services to peace and understanding between peoples"—a high demand. Internationally respected persons and institutions can be taken into consideration whose work corresponds to the ideals of Otto Hahn and in whom there is a close affinity of conviction and mind.

The Otto Hahn Peace Medal of 1988 was given to the former Italian State President Sandro Pertini "for outstanding services to peace and understanding between peoples, political morality, and practical humanity".

The Otto Hahn Peace Medal for 1989/90 was received by the Soviet State President Mikhail Gorbachev "for outstanding services... in atomic disarmament of the great powers and laying the ground for the political reorganisation of Europe".

---

[6] *'Deutsche Gesellschaft für die Vereinten Nationen'.*

The Otto Hahn Peace Medal for 1991 was awarded to Simon Wiesenthal, the founder and director of the Jewish Documentation Centre in Vienna, "in particular for his exemplary work for right and justice, dignity and tolerance, and reconciliation between men".

A eulogy for Mikhail Gorbachev, held on 25 November 1990, in Berlin, was given by the Swiss writer and dramatist Friedrich Dürrenmatt. From the play "Die Physiker" [7] and his essay on Albert Einstein for the latter's 100th anniversary of his birth, the writer seems to have been predestined as only a few others have been. The Gorbachev eulogy was his last literary performance. Three weeks later Dürrenmatt died, one of the great figures of Swiss contemporary literature.

What Dürrenmatt said at the end sounds like a legacy. He recognised Hahn's part in the changing of our outlook on the world and Gorbachev's contribution to the end of the nuclear arms race and the cold war. Dürrenmatt saw both inextricably linked with each other:

"... atomic weapons have been invented, and that can not be undone. What we need is the fearless common sense of Mikhail Gorbachev. What it can bring about we do not know... But fearless common sense is the only thing which we will have at our disposal in the future, perhaps even for our survival through it, ... the only thread to grasp at to pluck us from destruction".

Otto Hahn demonstrated this fearless common sense.

---

[7] *'The Physicist'.*

20
Timetable

1879	*8 March*. Otto Hahn born in Frankfurt am Main.
1895	*8 November*. Discovery of Röntgen rays.
1896	*January*. Röntgen published his work 'On a New Kind of Ray', which instituted a new epoch in scientific knowledge.
	1 March. Discovery of radioactivity by Bequerel.
1897	Otto Hahn begins studying chemistry at the Universities of Marburg and Münich (until 1901).
	Proof that the electron is a component of the atom.
1898	Marie and Pierre Curie establish the new radioactive elements polonium and radium.
1900	Discovery of the elementary quantum of action by Max Planck.
1901	*24 July*. Hahn obtains his Doctorate of Philosophy at the University of Marburg with his dissertation 'On Bromine Derivatives of Isoeugenols'.
1902	*1 October*. The start of a two year assistantship in the Institute of Chemistry at the University of Marburg with Professor Zincke.
	The development (until 1903) of the theory of radioactive decay by Rutherford and Soddy.
1903/04	The first speculations about atomic energy.
1904	*October*. Hahn travels to London for a nine month study visit with William Ramsay at University College, London.

1905 *16 March.* Ramsay makes known at The Royal Society Hahn's discovery of radio-thorium.

24 March. Hahn's first of over 250 scientific publications on radium research appears in *Proceedings of The Royal Society*.

September. Hahn travels to Montreal to Rutherford, and starts upon a nine month study visit at McGill University; he discovers the radioelements thorium C′ and radioactinium (1905/06).

Einstein proves the quantum nature of light and establishes his theory of relativity.

1906 *October.* The start of radioactive research work in the wooden workshop of the Institute of Chemistry of the University of Berlin.

1907 *9–12 May.* Hahn's lecture and contributions to discussions at the Bunsen Meeting on 'radioactivity' in Hamburg.

15 June. Habilitation at the Faculty of Philosophy of the University of Berlin.

Discovery of mesothorium.

October. The start of the thirty year long collaboration with the physicist Lise Meitner.

1908 *15 May.* The first of 50, in all, joint Hahn–Meitner publications appears.

Discovery of actinium C″.

1909 Explanation of the phenomenon of radioactive reactions;

The new radioelements radium C″ and thorium C″ are found.

August. Journey to Winnipeg (Canada) to a Meeting of the British Association.

1910 *13–15 September.* Participant at the International Radium Congress in Brussels. Member of the newly founded Radium Standard Commission.

1911 *11 January.* Foundation of the Kaiser Wilhelm Gesellschaft for the Furthering of the Sciences.

1912 *25–28 May.* Participant at the Meeting of the Radium Standard Commission in Paris.

August. Rutherford formulates his theory of the existence of the atomic nucleus after he had already spoken of this supposition in 1911.

23 October. The official opening of the Kaiser Wilhelm Institute for Chemistry in Berlin–Dahlem; Hahn becomes Director of a division for radioactive research.

1913 *22 March.* Marriage to Edith Junghans.

Soddy discovers the isotopes of the chemical elements.

New model of the atom according to Rutherford and Bohr.

1914   *1 August.* The start of the First World War; Hahn is conscripted into a territorial reserve regiment.

1915   *January.* Hahn is ordered to a gas combat unit led by Fritz Haber. *22 April.* The start of the gas war.

1916   *December.* Transfer of Hahn to the Supreme Headquarters in Berlin.

1918   *15 May.* The Hahn–Meitner publication about the discovery of the element number 91 (protoactinium).
November. Revolution in Germany. The fall of the monarchy.

1919   *June.* Rutherford's publication about the first artificial transformation of the atomic nucleus.

1921   *February.* Hahn's publication about the discovery of the isotope uranium Z (the proof of nuclear isomerism).
The start of work in the field of applied radiochemistry.
The bestowal of the title of Extraordinary Professor.

1924   *6 November.* Voted an Ordinary Member of the Prussian Academy of Sciences, Berlin.

1926   Otto Hahn's first scientific monograph *What Does Radio-Activity Teach Us about the History of the Earth?* appears from the Julius Springer Verlag, Berlin.

1928   Hahn becomes Director of the Kaiser Wilhelm Institute for Chemistry, provisionally led by him since 1926.

1932   Discovery of new elementary particles, of which the neutron gains special interest.
May. Bunsen Meeting on radioactivity, with the participation of prominent atomic researchers.

1933   *30 January.* The beginning of the National Socialists' power in Germany under Hitler.
February. Hahn's journey to the USA. Visiting professor at Cornell University, Ithaca, New York.
June. Premature return to Germany; Acting Director of the Kaiser Wilhelm Institute for Physical Chemistry and Electro-Chemistry after the resignation of Fritz Haber.
6 September. Lise Meitner's exclusion from the University of Berlin, withdrawal of the authority to teach on the order of the National Socialist Ministry.

1934   *15 January.* Publication of the discovery of artificial radioactivity by Irène and Frédéric Joliot-Curie.

31 January. To express his solidarity Hahn explains his resignation from the teaching staff of the University of Berlin.

June. Fermi and his coworkers make known the suspected discovery of the element 93, formed by the irradiation of uranium with neutrons.

September. Hahn and Meitner's journey to Moscow and Leningrad as participants in the Mendeleev Congress.

The start of work on the elucidation of the processes resulting from the irradiation of uranium by neutrons.

1935   *11 January.* Publication of the first Hahn–Meitner work on the uranium problem.

29 January. Hahn's address at the commemoration ceremony for Fritz Haber forbidden by the authorities.

2 August. The first Hahn–Meitner–Strassmann publication appears.

The mistaken discovery of the elements 93 to 97 (the 'false' trans-uranics) in the course of the years 1935–1938.

1936   *March.* The finding of the uranium isotope 239, which formed the first 'real' trans-uranium element 93, which, however, Hahn did not seek.

Hahn's textbook *Applied Radio-Chemistry* is published in Ithaca, New York, Oxford, and London; a Russian translation is soon published.

1938   *13 July.* Lise Meitner leaves Germany.

18 November. The last 'false' work on trans-uranium by Hahn–Strassmann appears.

17 December. Hahn and Strassmann succeed in the chemical proof of uranium fission.

22 December. Completion and submission of the manuscript about the discovery of uranium fission.

1939   *6 January.* Publication of the first Hahn–Strassmann work on uranium fission in *Naturwissenschaften*, published by Julius Springer Verlag, Berlin.

26 January. Bohr tells of the fission of uranium at a physics congress in Washington.

11 February. Meitner and Frisch publish their theoretical explanation of uranium fission.

8 March. Hahn's 60th birthday.

18 March. The Joliot-Curies' publication about the proof of additional fission neutrons. The possibility of a chain reaction becomes real.

June/August. Publications by Siegfried Flügge about the technological exploitation of atomic energy.

2 August. Einstein recommends to the US President Roosevelt the starting of work on the production of the atom bomb, in order to beat Hitler.

1 September. The start of the Second World War.

The German atomic research programme is run up, plans for a 'uranium machine'.

1940   The start of the work on the American atom bomb project.

1941   *August.* Houtermans' secret report about the triggering of nuclear chain reactions.

1942   *26 & 27 February.* Hahn's participation in a secret session of the Nuclear Physics Working Group (Reich's Research Department / Army Weapons Branch).

4 June. Hahn takes part in a conference under Reich's Minister Speer about the continuation of the atomic research programme.

2 December. Fermi succeeds in the USA in setting into action the first self-sustaining chain reaction (the 'uranium machine').

1944   *11 February.* Destruction of the Kaiser Wilhelm Institute for Chemistry during a bombing raid.

The removal of the Institute to Tailfingen.

1945   *25 April.* Hahn and nine other atomic researchers are taken prisoner by an Allies' commando operation.

8 May. The unconditional surrender of Germany.

3 July. The arrival of the internees at the English country seat of Farm Hall, in the general vicinity of Cambridge.

6 & 9 August. The dropping of the two American atom bombs on Hiroshima and Nagasaki.

15 November. Announcement of the award of the Nobel Prize to Otto Hahn.

1946   *3 January.* Return of the interned atomic researchers.

23 & 24 January. Hahn takes part in a Röntgen Jubilee celebration in Hamburg. The first interview about the German work on atomic energy during the war.

1 April. Hahn becomes President of the Kaiser Wilhelm Gesellschaft.

10 December. Receipt of the Nobel Prize in Stockholm.

1948   *26 February.* Foundation of the Max Planck Gesellschaft out of the Kaiser Wilhelm Gesellschaft. Otto Hahn becomes its President.

1949   *8 March.* 70th birthday.

23 September. Announcement of the first Soviet atom bomb.

7 September. Foundation of the Bundesrepublic of Germany.

12 December. Chancellor Adenauer discusses questions of future research policy with Hahn and Heisenberg.

1950   *March.* The first Stockholm appeal for the banning of atomic weapons.

1952   *1 November.* Explosion of the first American hydrogen bomb.

1953   *12 August.* Explosion of the first Soviet hydrogen bomb.

1954   *1 March.* Detonation of the first American three stage bomb.

8 March. 75th birthday.

27 June. The first atomic power station in the world starts its operation in Obninsk (USSR).

1955   *13 February.* Hahn's radio appeal about 'cobalt 60'—a danger or a blessing for the world.

15 July. The Mainau Rally of Nobel Prize winners, proposed by Hahn, appeals to the world's statesmen to refrain from force as the means of politics.

July/August. International Atomic Conferences on the peaceful application of atomic energy in Moscow and Geneva. Hahn leads the Geneva delegation.

September. The first ever award of the Otto Hahn Prize for Chemistry and Physics, to Lise Meitner, amongst others.

November/December. Hahn's journey with his son Hanno in the USA; lectures and tours of research sites.

1956   *January.* Founding assembly of the German Atomic Commission. Strauss is elected Chairman, and Hahn Vice President.

19 November. Hahn and eleven other scientists forewarn Strauss by letter about equipping the German armed forces with atomic weapons.

1957   *13 March.* Announcement from NATO Headquarters that tactical atomic weapons will be stored in the BRD.

5 April. Adenauer minimises atomic weapons as 'matured artillery'.

12 April. Publication of the Göttingen Declaration of eighteen atomic researchers against the atomic arming of the Bundesrepublic.

17 April. Summons of Hahn and four other atomic researchers to the Chancellor's office.

13 November. Hahn's address in Vienna, 'Atomic Energy for Peace of for War?'

5 December. Launch of the first atomic ship in the world for civilian purposes, the Soviet ice-breaker 'Lenin'.

1958   *15 January.* Announcement of the Pauling appeal endorsed by Hahn; demand for the settling of the agreement about the use of nuclear weapons tests.

24 & 25 April. Hahn as the guest of the celebratory meeting of the German Academy of Sciences in East Berlin in honour of the 100th birthday of Max Planck.

30 September. Hahn's address 'On the History of Uranium Fission and the Consequences Flowing from It'.

1959   *8 March.* 80th birthday.

14 March. Official opening of the Hahn–Meitner Institute for Nuclear Research in West Berlin, in the presence of those it was named after.

November/December. Hahn's journey to Israel is the start of a German–Israeli collaboration in the sphere of science.

1960   *19 May.* Hahn gives up his office as President of the Max Planck Gesellschaft to Butenandt and becomes Honorary President.

29 August. His son Hanno and daughter in law Ilse killed in an accident.

1961   *January.* Hahn signs Pauling's appeal 'No More Atomic States'.

1962   Publication of Hahn's autobiography *From Radio-Thorium to Uranium Fission.*

1963   *5 August.* Treaty for the cessation of nuclear weapons testing in the atmosphere and under water. Hahn welcomes the agreement.

1964   *13 June.* Participant in the launch of the first European atomic powered merchant ship, the 'Otto Hahn'.

1966	*June.* Journey to Czechoslovakia, to the Joachimsthal radium spa.

1966   *June.* Journey to Czechoslovakia, to the Joachimsthal radium spa.

November. The unveiling of a commemorative plaque for the researches of Otto Hahn and Lise Meitner in the former Institute of Chemistry of the University of Berlin.

1968   *1 July.* Settling of the treaty on the embargo on nuclear weapons.

28 July. Otto Hahn dies in Göttingen after a long stay in a clinic.

14 August. His wife, Edith Hahn, dies.

August. Otto Hahn's autobiography *My Life* appears.

27 October. Lise Meitner dies in Cambridge.

An antarctic island and a crater on the Moon are named after Otto Hahn.

1969   Establishment of the Otto Hahn Foundation of the City of Frankfurt am Main, the Otto Hahn Prize for outstanding achievements in Hahn's field of scientific work and the peaceful use of nuclear energy.

1970   The first account of the chemical element 105, later named Hahnium.

1979   *8 March.* 100th birthday with celebratory events.

1980   *22 April.* Fritz Strassmann dies.

1982   The first account of the element 109, later named Meitnerium.

1988   *17 December.* 50th anniversary of the discovery of uranium fission.

Endowment of the gold Otto Hahn Peace Medal for outstanding services to peace, disarmament, and international understanding. The first Prize Winner is the former Italian State President, Pertini.

Militant opponents of nuclear power destroy an exhibition in Berlin '50 Years of the Discovery of Nuclear Fission'.

1990   The presentation of the Otto Hahn Peace Prize awarded in 1989 to the Soviet Union State President, Gorbachev. Eulogy by the writer Dürrenmatt.

1991   *17 December.* Award of the Otto Hahn Peace Medal to Simon Wiesenthal, Director of the Jewish Documentation Centre in Vienna.

1992   *8 March.* After the reform of the statutes the Otto Hahn Prize of the City of Frankfurt am Main is awarded to the physician Olga Aleinikova of Minsk in recognition of her humanitarian dedication to the radiation damaged children of the Chernobyl catastrophe.

21
Sources and Pointers

It will be understood that in the confines of this book it has not been possible to cite the over 3,500 sources which were recorded, analysed, and evaluated for the present biography, which was first published in 1978 and now is virtually a new work. This source material embraces specialist scientific journals and monographs both in and outside Germany (primarily from 1896 to 1944), literature on the history of the times and the history of science as well as contemporary, further information which was only to be gained from daily papers and periodicals of the years 1910 to 1912, 1930 to 1944, and from 1945 to the present day, and further material from archives and personal effects. From the abundance of primary and secondary bibliographical literature used, only the most important publications up to the present day have been cited, ordered by the year of publication.

Amongst those who have contributed to the success of the work the first place goes to Dietrich Hahn of Ottobrunn, grandson and only descendant of Otto Hahn, and Dr. Rainer Stumpe of Springer–Verlag. They are due my thanks, as well as the President's office of the Max Planck Gesellschaft, München, and the Archive of the History of the Max Planck Gesellschaft, Berlin–Dahlem, for permission to use the material from Otto Hahn's estate, in addition to Marie-Louise Rehder of Göttingen, the secretary of Professor Hahn for many years, who later fulfilled all my requests for material and helped to answer detailed questions.

For personal advice, especially on the history of the discovery of fission of the nucleus and on the history of the exploitation of nuclear energy, I am obligated with much thanks to Professor Dr. Fritz Strassmann of Mainz,

Hahn's coworker for many years, and to his wife Irmgard Strassmann, as well as to Professor Dr. (H.C.) Manfred von Ardenne of Dresden.

Otto Hahn (1962). *Vom Radiothor zur Uranspaltung, Eine wissenschaftliche Selbstbiographie.* (Braunschweig); Reprint (1988).

Friedrich Herneck (1965). *Bahnbrecher des Atomzeitalters,* 9th edn., (Berlin), (1984).

Otto Hahn, (1986). *Mein Leben,* (Münich); newly edited by Dietrich Hahn, (Münich) (1986).

Walther Gerlach (1969). *Otto Hahn. Ein Forscherleben unserer Zeit,* (Münich).

Ernst Berninger (1969). *Otto Hahn. Eine Bilddokumentation,* (Münich).

Ernst Berninger (1974). *Otto Hahn in Selbstzeugnissen and Bilddokumenten,* (Reinbek bei Hamburg).

Dietrich Hahn (ed.) (1975). *Otto Hahn. Erlebnisse and Erkenntnisse,* (Düsseldorf and Vienna); containing *Otto Hahn. Erinnerungen 1901–1945.*

Jost Herbig (1976). *Kettenreaktion. Das Drama der Atomphysiker.* (Münich, Vienna).

Armin Hermann (1977). *Die Jahrhundertwissenschaft. Werner Heisenberg und die Physik seiner Zeit.* (Stuttgart); new edn. (1993), *Die Jahrhundertwissenschaft. Werner Heisenberg und die Geschichte der Atomphysik,* (Reinbek bei Hamburg).

Klaus Hoffmann (1978). *Otto Hahn. Stationen aus dem Leben eines Atomforschers. Biographie,* (Berlin); 4[th]edn. (1987), with a foreword by Manfred von Ardenne.

Dietrich Hahn (1979). *Otto Hahn—Begründer des Atomzeitalters, Eine Biographie in Bildern und Dokumenten,* (Münich).

Horst Wohlfarth (ed) (1979). *40 Jahre Kernspaltung. Eine Einführung in die Originalliteratur,* (Darmstadt).

Otto Robert Frisch (1981). *Woran ich mich erinnere. Physik und Physiker meiner Zeit,* (Stuttgart).

Dietrich Hahn (1981). *Otto Hahn in der Kritik*, (Münich).

Fritz Krafft (1981). *Im Schatten der Sensation. Leben und Wirken von Fritz Strassmann*, (Weinheim; Deerfield Beach; Basel).

Walther Gerlach and Dietrich Hahn (1984). *Otto Hahn. Ein Forscherleben unserer Zeit*, (Stuttgart).

Charlotte Kerner (1986). *Lise, Atomphysikerin. Die Lebensgeschichte der Lise Meitner*, (Weinheim und Basel).

Helmut J. Fischer (1987). *Hitler und die Atombombe. Bericht eines Zeitzeugen*, (Asendorf).

Arnold Kramish (1987). *Der Greif. Paul Rosbaud—der Mann, der Hitlers Atompläne scheitern ließ*, (Münich).

Dietrich Hahn (ed) (1986). *Otto Hahn. Leben und Werk in Texten und Bildern*, (Frankfurt am Main).

Jost Lemmerich (1988). *Die Geschichte der Entdeckung der Kernspaltung. Katalog zur Ausstellung*, (Berlin).

Richard Rhodes (1988). *Die Atombombe oder Die Geschichte des 8. Schöpfungstages*, (Nordlingen).

Carl Friedrich von Weizsäcker (1988). *Bewußtseinswandel*, (Münich; Vienna).

Rudolf Heinrich and Hans-Reinhard Bachmann (1989). *Walther Gerlach. Physiker—Lehrer—Organisator. Dokumente aus seinem Nachlaß*, (Münich).

Alwyn McKay (1989). *Das Atomzeitalter. Von den Anfängen zur Gegenwart*, (Berlin, Heidelberg, New York).

Patricia Rife (1990). *Lise Meitner. Ein Leben für die Wissenschaft*, (Düsseldorf).

Rudolf Vierhaus and Bernhard von Brocke (eds) (1990). *Forschung im Spannungsfeld von Politik und Gesellschaft. Geschichte und Struktur der Kaiser-Wilhelm-/Max-Planck-Gesellschaft*, (Stuttgart).

Mark Walker (1990). *Die Uranmaschine. Mythos und Wirklichkeit der deutschen Atombombe*, (Berlin).

Wolfgang Menge (1991)
Ende der Unschuld. Die Deutschen und ihre Atombombe, (Berlin).

Thomas Powers (1993)
Heisenbergs Krieg. Die Geheimgeschichte der deutschen Atombombe, (Hamburg).

Translator's Notes on the Text

NOTE 1: (See page 7):
'Farm Hall Report No. 4' records remarks akin to this one sentence, but not the same, and as being made in a different conversation not at this juncture.

NOTE 2: (See page 7):
This remark is not recorded in the 'Farm Hall Report No. 4'.

NOTE 3: (See page 7):
This sequence of events described differs from what is recorded in 'Farm Hall Report No. 4'. See Note 5 following.

NOTE 4: (See page 8):
Because of the release of the Farm Hall Transcripts some years after he wrote this book, the author could make no mention that it was Gerlach whom the Farm Hall Report of *Operation Epsilon* recorded as being feared potentially suicidal. The Report records that Gerlach went up to his room after a remark by Korsching, and was heard to be sobbing. Neither could the author mention that the Report also records that it was von Laue and Harteck who were the first to go up to Gerlach in his room to soothe him. *See* (1993), 'Farm Hall Report No. 4', *Operation Epsilon*, (Institute of Physics Publishing: Bristol, and Philadelphia, PA), p. 79; Jeremy Bernstein (1996) 'Farm Hall Report No. 4', *Hitler's Uranium Club*, (American Institute of Physics Press: Woodbury, NY), p. 133.

A paragraph or two later in 'Farm Hall Report No. 4', after reporting that Gerlach said to von Laue and Harteck "Please leave me alone", it is mentioned that it was a little later that Hahn went up to comfort Gerlach, and a record of their conversation is given, which showed Gerlach to have become responsive through their colloquy. After three exchanges between the two Harteck comes in (obviously once again) and a detailed conversation develops with Gerlach taking an active part, and which lasts for some time. Gerlach clearly recovered his composure significantly with that conversation. There is no further mention of worries about Gerlach that night.

After that conversation the Report relates that Hahn and Heisenberg discussed its "matter(s raised within that conversation) alone together. Hahn explained to Heisenberg that he was himself very upset about the whole thing. He said he could not really understand why Gerlach had taken it so badly" (*See* Item 7 for 6 August 1945, Farm Hall Report No. 4). This direct record of the contributions of all the parties to these conversations counters the contention that Hahn was of any suicidal mind, let alone sufficient to cause his colleagues serious concern, and reveals that it was Gerlach who was badly affected with such a mood, here incorrectly imprecated upon Hahn.

NOTE 5: (See page 8):
There is no record in 'Farm Hall Report No. 4' of such sentiments about Hahn being expressed by anybody. See Note 4 preceding.

NOTE 6: (See page 8):
There is no record of the little episode described here extant in the Farm Hall Report. This in itself is significant because of the intensive work undertaken of installing microphones everywhere possible, and a very keen listening watch kept (as ordered) to discover the reactions, comments, and behaviour of all the internees immediately after the news of the dropping of the first atom bomb.

In contrast, the Farm Hall Report shows a very clear statement that Gerlach noticeably showed signs of being potentially suicidal, and that von Laue and Harteck actually went to his room "early in the evening" to quieten him. The Report has no record of Hahn being potentially suicidal, although it is clearly stated that he was very deeply affected.

Bagge's diary states that von Laue came to his room at 2 a.m., and that together they listened in Bagge's room to Hahn, next door, through the wall—and not going to Hahn's door and looking in—until Hahn was clearly asleep. The author also drew his information from Gerlach's diary.

NOTE 7: (See page 31):
Henri Poincaré's principal claim to fame was as a mathematician. Amongst his visionary concepts he produced ideas that came to form the basis of algebraic topology.

NOTE 8: (See pages 46, 66):
'Geheimrat' and 'Geheime Kommerzienrat'. 'Geheim', literally 'secret', corresponds to advisor or 'Privy', and the ending 'Rat' corresponds to 'Councillor'. Thus 'Geheimrat' corresponds to the honorary title of 'Privy Councillor'. 'Kommerzienrat' is an honorary title conferred on financiers and business magnates. 'Geheime Kommerzienrat' thus corresponds, but only broadly, to a notion such as 'Privy Commercial Councillor' which might exist in a monarchical society.

NOTE 9: (See page 100):
The actual experiment did not use a cyclotron but a Heath Robinson apparatus which was actually a linear accelerator. The very high accelerating voltage was produced by a newly invented system of rectifiers and capacitors—subsequently

nicknamed the 'Cockroft and Walton' multiplier after the inventive pair, and often used over the next two decades in electronic equipment requiring extra high tension at low power—which produced a final DC voltage which was, indeed, an integer multiple of the input AC voltage corresponding to the number of 'multiplier' stages. The voltage was applied across the source–target pair, each mounted at either end of an evacuated tube consisting of two cylindrical glass vessels used in the petrol pumps of that day, sealed to the end plates and each other by children's plasticine. Rutherford was asked if he had a motor available to drive the vacuum pump they had found, to which he replied that he thought there was one in a cupboard full of equipment brought back from the fateful and fatal antarctic expedition to the South Pole of Captain Scott. They did indeed find one, which Scott had taken—along with many other mechanical artifacts— to test their durability in extreme cold. The experiment of Cockroft and Walton was notable for these many remarkable, and little known, facets, and continued in an exemplary manner the English tradition of obtaining profound results and discoveries with the simplest and (even) crudest of equipment.

These unexpected details were given by Sir John Cockroft in an invited lecture to a university physics society attended by the translator when a student.

NOTE 10: (See page 196):
The German surrender took place at Lüneburg on May 4, 1945, and not in Rheims on May 7. On that latter date the party of which Hahn was a member, and which consisted of von Laue, von Weizsäcker, Wirtz, Bagge, and Korsching, was already in Rheims—having being detained at 75, Rue Gambetta since May 2—and it was at 5 p.m. that the party was flown by Dakota to the Paris area and then quartered in the old Chateau du Chesnay (known as 'Dustbin' to the Allies) in Versailles.

It is just possible that Hahn and his party at Rheims heard of the end of the war before they flew to Paris, as the first official news of the end of the war was broadcast on the Six O'Clock News of the BBC Home Service that same evening, the first news broadcast since the One O'Clock News. 6 p.m. in England may well have been 6 p.m. in Rheims, instead of 7 p.m., since Britain ran 'double summer time' during the war. Therefore the news of the end of the war may well have been able to have been given to the party in Rheims by the British military before their departure for Paris. It was two days later, on May 9 that Heisenberg and Diebner were added to the party at 'Dustbin'. (See, for example, (1993), *Operation Epsilon*, (IOP Publishing: Bristol, UK, and Philadelphia, PA), pp. 19–21).

NOTE 11: (See page 196):
Hahn was having fun with the official phrase 'detained *at* His Majesty's pleasure', wording always used in sentences of the court passed on persons considered a grave danger to society (such as criminal lunatics or very young murderers) and whose period of recovery to a condition safe to the population at large can only be considered as completely unquantifiable. It is likely that from his good acquaintances made in England Hahn understood all of this; and one might speculate that with his puckish sense of humour he may even have been quietly enjoying the knowledge that his guard might have had no idea at all that he also fully understood the inference of this phrase that they were dangerous lunatics only fit for being locked up.

NOTE 12: (See page 230):

The original text names this officer as 'Percy R. Earl of Bandon'. The earldom was of the Irish peerage, and the family name of old was Bernard. At the time of this incident Air Marshall the Earl Bandon was C-in-C of the 2nd Tactical Air Force (1955–57). He was awarded the DSO in 1940, was thrice mentioned in despatches, and was awarded the American DFC and Bronze Star. He was AOC No. 2 Group BAFO Germany 1950–51, and of No. 11 Group, 1951–53.

NOTE 13: (See page 230):

Air Chief Marshall the Earl Bandon's military career actually continued at a high level. He was appointed C-in-C of the Far East Air Force of the Royal Air Force, 1957–60. Following that he was appointed to the very senior post of Commander of the Allied Forces Central Europe, 1961–63.

A much more likely explanation of his revelation of a probable 'secret'—and especially in view of his previous experience of air command in Germany—is that he was given a briefing for speaking to the press in which he had been reassured that such sensitive information was approved for release. He would undoubtedly have checked beforehand the official public relations channels (through others of which at least one instance of a similar blunder occurred a few years later) to establish that it was agreed by superiors that the information could be released.

NOTE 14: (See page 235):

The author used the word 'humanist' in the original text, but Schweitzer's considerable works as a theologian before turning to medicine and missionary work can only indicate 'humanitarian' as the true meaning.

Index